About Island Press

Island Press is the only nonprofit organization in the United States whose principal purpose is the publication of books on environmental issues and natural resource management. We provide solutions-oriented information to professionals, public officials, business and community leaders, and concerned citizens who are shaping responses to environmental problems.

In 2001, Island Press celebrates its seventeenth anniversary as the leading provider of timely and practical books that take a multidisciplinary approach to critical environmental concerns. Our growing list of titles reflects our commitment to bringing the best of an expanding body of literature to the environmental community throughout North America and the world.

Support for Island Press is provided by The Bullitt Foundation, The Mary Flagler Cary Charitable Trust, The Nathan Cummings Foundation, Geraldine R. Dodge Foundation, Doris Duke Charitable Foundation, The Charles Engelhard Foundation, The Ford Foundation, The George Gund Foundation, The Vira I. Heinz Endowment, The William and Flora Hewlett Foundation, W. Alton Jones Foundation, The John D. and Catherine T. MacArthur Foundation, The Andrew W. Mellon Foundation, The Charles Stewart Mott Foundation, The Curtis and Edith Munson Foundation, National Fish and Wildlife Foundation, The New-Land Foundation, Oak Foundation, The Overbrook Foundation, The David and Lucile Packard Foundation, The Pew Charitable Trusts, Rockefeller Brothers Fund, The Winslow Foundation, and other generous donors.

About The Nature Conservancy

The mission of The Nature Conservancy is to preserve plants, animals, and natural communities that represent the diversity of life on Earth by promoting the lands and waters they need to survive.

Since 1951, the Conservancy, a nonprofit membership organization headquartered in Arlington, Virginia, has worked with local conservationists both in the United States and internationally to identify and protect critical habitat. As a global organization, The Nature Conservancy has been instrumental in protecting more than 92 million acres of land in 29 countries around the world.

The Conservancy recognizes that local organizations have the best insight into local conservation issues. It therefore works to support partners, ensuring that they have the resources and skills to make decisions that will guarantee a rich natural legacy.

For more information about The Nature Conservancy and its work, visit its Web site at www.tnc.org.

DESIGNING
FIELD STUDIES FOR
BIODIVERSITY
CONSERVATION

Designing Field Studies for Biodiversity Conservation

Peter Feinsinger

The Nature Conservancy

ISLAND PRESS
Washington • Covelo • London

Library of Congress Cataloging-in-Publication Data

Feinsinger, Peter, 1948–
  Designing field studies for biodiversity conservation / Peter
Feinsinger.
     p.   cm.
Includes bibliographical references (p.   ).
  ISBN 1-55963-878-8 (pbk. : alk. paper)
  1. Biological diversity conservation. I. Title.
  QH75 .F45 2001
  333.95'16—dc21   2001001682

British Cataloguing in Pubication data available.

Printed on recycled, acid-free paper ♲

Manufactured in the United States of America
10 9 8 7 6 5 4 3 2 1

To the memory of Archie F. Carr Jr. (1909–87), who reminded us gently that common sense and a knowledge of natural history are far more potent conservation tools than anything labeled with an acronym

Contents

List of Figures, Tables, and Boxes

Figures

Tables

Boxes

Preface

When Kent Redford, then at The Nature Conservancy, invited me to write this book for the Conservancy's Latin America and Caribbean Division, my reaction alternated between wild enthusiasm and mild trepidation. Enthusiasm, because the philosophy of the book that Kent was proposing coincided with my personal belief in the urgency of fostering self-sufficiency and intellectual independence among the diverse array of Latin American professionals involved directly or indirectly with biodiversity conservation. Trepidation, for I knew (and still know) next to nothing about the politics and policy strictures of conservation and management, which implied that many of the suggestions I'd make or tangents I'd take in such a text would strike many readers as being hopelessly impractical, irrelevant, or naïve. Soon afterwards, though, a presentation of some preliminary thoughts during a workshop at the Conservancy's III Semana Conservacionista in Quito evoked an unexpectedly positive response. Thus encouraged, I squelched the misgivings and forged ahead.

In part, then, this text follows the scheme that was presented verbally to a patient audience of conservation professionals, from the Conservancy's partner organizations throughout Latin American and the Caribbean, who came to Quito in May 1995. More fundamentally, though, the book derives from a much broader base of experiences with building local capacity in "hands-on and minds-on science" in a number of settings in South America and elsewhere. At various times these courses and workshops have included rural school teachers, university students and professors in ecology, members of indigenous communities, professional conservation biologists, park guards, protected area managers, and personnel of local nongovernmental conservation organizations. The book's points of view also result from thirty years' worth of personal "hands-on science," ecological field research I've undertaken in forested, agricultural, and semi-desert landscapes of Central and South America and the Caribbean.

Therefore, the book's intended audience is an inclusive one. You, the reader, probably aren't a Latin American elementary school teacher, member of a rural community, or park guard (although see chapter 10). You might be, though, a conservation ecologist, wildlife biologist, or other professional working on specific projects in or around a protected area; the manager or conservation science specialist of that protected area; a graduate or undergraduate student in ecology, conservation biology, natural forest management, wildlife management, environmental science, or a related field; a researcher at an institute of agro-ecology or sustainable forestry—in short, anyone whose concerns touch on the broad themes of "biodiversity" and "biodiversity conservation."

The book will focus on questions pertinent to conservation and management, in and around protected areas. Don't turn away, though, if you're a student proposing a thesis project in basic field ecology—or someone whose concern is the "semi-natural matrix" (Brown, Curtin, and Braithwaite 2001) far from any protected area. The principles are the same. If you live in the North Temperate Zone, Africa, Asia, or Australia, don't be put off by the emphasis on Latin America or the fact that translations into Spanish and Portuguese will appear soon. A number of citations in the text, if not always the text's contents, refer to the North Temperate nations and Australia. Again, the principles are the same. Don't expect this book, though, to present a thorough treatise on the field of conservation biology. Turn to Meffe and Carroll (1997) or Primack et al. (2001) for that. Don't expect the book to deal with the critically important social dimensions of conservation or (except for parts of chapters 7 and 10) the most effective means of working in collaboration with local communities and other "stakeholders." Margoluis and Salafsky (1998) and many other books already do a great job of this. Finally, don't expect this book to provide a practical guide to planning a complete, multidisciplinary, regional conservation strategy. The Nature Conservancy's "Site Conservation Manuals" do just that.[1]

Instead, this book's intent is to convince you that scientific inquiry, as a means of asking and answering questions about your surroundings, is indispensable to your goals whether those be thesis projects in field ecology or conservation guidelines regarding natural forest management. The concurrent purpose is to dispel at least five common misconceptions:

- the perception that scientific inquiry is just a sterile, self-serving, narcissistic pastime of the Ivory Tower that has no relevance to the real world of conservation and management;

- the belief that the trappings of science (for example, high-tech instruments and approaches, sophisticated and incomprehensible language, statistical magic tricks) are what constitute scientific inquiry and that therefore scientific inquiry is to be either feared or spurned;

- the complementary assumption that certain other trappings of science (e.g., indices of species diversity, canned programs in statistical analysis, terms such as *indicator* and *keystone* and even *biodiversity*) can be borrowed for conservation purposes without having to reflect profoundly on their hidden assumptions or on whether they make sense biologically;

- the belief that objective decisions on strategies for biodiversity conservation, inside or outside of protected areas, can be made without a thorough understanding of scientific inquiry;

- the idea that use of scientific inquiry requires advanced training and consequently is inaccessible to "ordinary" people, if not to many conservation professionals.

To practice what it preaches, then, this book will *not* aim to impress, intimidate, or overwhelm you with jargon. It lacks glowing descriptions of space-age technology or statistical packages that you simply must acquire. You'll find no elegant mathematical models. If you're seeking a diplomatic, scholarly review that encompasses an awe-inspiring number of references to a vast body of erudite literature, you'll be disappointed. Neither will the book provide you with "cookbook recipes" or paternalistically proffered guidelines to be followed without question. Instead, I hope that the book will catalyze thought, will goad you to reflect broadly and deeply while going through the process of developing conservation guidelines or thesis projects, and will convince you of the reasonableness of "doing science" in order to further goals in management, conservation, and basic field biology.

A final note, on the book's structure. "Boxes" inserted into the text present exercises crucial to building your expertise in the inquiry process and applying that to your particular goals, or else they present important philosophical and practical points that are outside the main flow of the text. Most chapter notes (grouped at the end of the text) present lists of citations for particular ideas or techniques mentioned in the text, and sometimes evaluate those. Other chapter notes present the historical background of, explain, or comment upon particular points made in the text. Appendices present details on quantitative methods, lists of resources including human resources, and (appendix D) a succinct approach to understanding design and statistics.

Acknowledgments

I take full responsibility for the frequent strong opinions expressed in the book, as well as for any misconceptions, misrepresentations, or other errors of commission and omission. Much of the advice you'll encounter has been obtained the hard way in thirty-plus years of learning from my own mistakes (an ongoing process); I fervently hope that no reader will ever analyze my published studies with this book in hand. Nevertheless, I owe a tremendous debt to numerous people with whom I've had the honor to interact, from whom I've learned and continue to learn, and who have greatly influenced the points of view presented here. Long ago, my mentors Richard Root and Robert Whittaker cautioned me against ignoring the complexities of natural history during the search for ecological patterns. Faculty colleagues, postdoctoral fellows, and graduate students at the University of Florida molded my thinking during the sixteen years I spent on the faculty there. The most permeating and durable of those influences came from Marty Crump, Jack Ewel, Alejandro Grajal, Martha Groom, Bert Klein, Gary Meffe, Carolina Murcia, Reed Noss, Jack Putz, Kent Redford, Carla Restrepo, John Robinson, Kirsten Silvius, and of course the late Archie Carr, who would have written about the book's themes with much greater eloquence and much less bombast. Most recently, the Department of Biological Sciences at Northern Arizona University has graciously tolerated an adjunct faculty member who's usually several thousand kilometers away or else in his study at home trying to write this book, while the Wildlife Conservation Society has graciously tolerated a maverick collaborator who has never visited the New York headquarters.

The proposal that scientific inquiry should be accessible to every person, as a tool for asking and answering questions about her or his surroundings, reflects many years of working on the "school-yard ecology" concept along with North American friends (especially Alan Berkowitz, Carol Brewer, Alejandro Grajal, Karen Hollweg, and Maria Minno) and a great number of South American friends from Venezuela to southern Chile (especially Samara Álvarez, Natalia Arango, Geovana Carreño, Andrea Caselli, María Elfi Chaves, Ruth Espinosa, Iván Flores, Paola del Giorgio, Emma Elgueta, Humberto Gómez, Margarita Herbel, Laura Lojano, Laura Margutti, Patricia Morellata, Isabel Novo, Lily Oviedo, Beatriz Parra, Amalia Pereda, Edmundo Rivera, Alejandra Roldán, Ricardo

Rozzi, Klaus Schütze Páez, and Rob Wallace). Also in Chile, Fabián Jáksic, Ricardo Rozzi, and Javier Simonetti spent a decade or more patiently introducing me to the "Southern Cone perspective" and weaning me from exclusive dependence on tropical landscapes, while in Argentina I have learned a great deal from personnel of Parques Nacionales in San Carlos de Bariloche, especially Laura Margutti and Juan Salguero; from my colleague Marcelo Aizen; and from Alejandro Brown and the research group of the Laboratorio de Investigaciones Ecológicas de las Yungas.

In Ecuador, colleagues at Fundación EcoCiencia (Quito), especially Rocío Alarcón, taught me a great deal about conservation and management at the local level. Jody Stallings of Proyecto SUBIR provided many opportunities for collaborating on applied research projects, for learning firsthand about the interaction between local communities and protected areas, and for developing scientific inquiries meaningful to those communities as well as to the conservation professionals involved. I'm grateful to Alejandro Grajal, in his former role at Wildlife Conservation Society, for supporting applied research and training activities under and out from under the SUBIR umbrella. Andrew Taber, Alejandro's successor at WCS, has diverted me to Bolivia, where work with WCS projects in the Gran Chaco (Andy Noss) and the Madidi-Pilón Lajas (Rob Wallace) regions has opened up entirely new perspectives. Thanks to Cristián Samper, in Colombia both the Fundación para la Educación Superior and the Instituto von Humboldt have supported courses, workshops, and training activities in conservation ecology and education. I thank the participants in the several Ecuadorian, Bolivian, and Colombian versions of the course "Diseño de Investigaciónes en Ecología de la Conservación" (Design of Field Research in Conservation Ecology) and my co-instructor in the Colombian versions, Carolina Murcia, for their infectious enthusiasm and thoughtful suggestions about the themes and details of this book. Participants in short versions of the "Diseño" course (Mexico, Venezuela, and Brazil) also provided keen insights, especially regarding the material in chapters 4 and 6.

Approaches to biodiversity conservation in the Old World prod us to reevaluate our New World perspectives. I thank Jackson Mutebi and other participants in workshops with communities around Bwindi Impenetrable National Park, Uganda, for their top-notch prodding during the time we worked together. I'm especially gratefully for the opportunity to work with park warden Paul-Ross Wagaba, a great hope for positive interactions between communities and the park who, not long after the workshops, was murdered by Rwandan infiltrators.

Finally, Kent Redford and Jennifer Shopland, formerly of The Nature Conservancy, provided encouragement and advice during the early stages of writing. Marty Crump carefully reviewed and edited several drafts, making a valiant if ultimately unsuccessful effort to tone down the rhetoric. Many colleagues kindly agreed to serve as resources (appendix C). Stuart Hurlbert and Douglas Johnson provided particularly expert reviews of those sections dealing with study design and statistics; they cannot be blamed if I sometimes failed to follow their sage advice. Jim Rieger at The Nature Conservancy thoughtfully reviewed the book from beginning to end, while Barbara Dean, Barbara Youngblood, and reviewers at Island Press provided many timely editorial suggestions. Rob Feinsinger drew the graphics, thereby greatly increasing the holdings of his coin collection. Universal Press Syndicate declined to grant permission to use Gary Larson's "Far Side" cartoons, thereby greatly reducing the probability that you'll truly enjoy reading this book.

CHAPTER 1

Introduction: What's Science Got to Do with It?

Despite the potential of applied ecology, there is still disagreement about the extent to which ecological science is applicable to real-world problems.

—Alicia Castillo and Victor M. Toledo (2000)

This book is intended for all those who work toward sustainable and sustained conservation of the landscapes that surround them along with the native biota those landscapes support. What does *conservation* mean, though? It seems that each of us has a unique and constantly changing definition. At this moment my own definition of conservation is *the field of study and action that concerns the management of the landscape so as (1) in the short and medium term, to minimize or buffer negative effects of human beings on nature, which includes the landscape's human inhabitants ourselves, and (2) in the long term, to provide other living beings with the maximum number of alternatives for tolerating and surviving our species' brief presence on this planet.*[1]

Getting at Conservation

How might conservation be achieved? The effects—positive, negative, and neutral—of humans on landscapes are the cumulative result of the individual choices that persons and institutions make. Perhaps sustainable and sustained conservation can be achieved only through education at all levels of society so that today's children, tomorrow's adults, become familiar with their natural surroundings, recognize the consequences that alternative decisions might have on those surroundings, and make

1

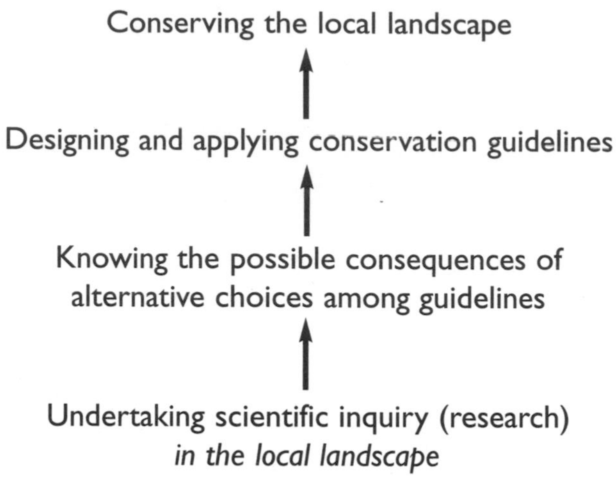

Conserving the local landscape

↑

Designing and applying conservation guidelines

↑

Knowing the possible consequences of
alternative choices among guidelines

↑

Undertaking scientific inquiry (research)
in the local landscape

Figure 1.1.
Why scientific inquiry should play a role in conservation and management.

decisions thoughtfully (Feinsinger, Mangutti, and Oviedo 1997; and see chapter 10). While we strive for that distant goal, though, management by conservation professionals, in consultation with local communities, provides one practical approach to biodiversity conservation (figure 1.1).[2]

How should such management be carried out? The people responsible must develop practical guidelines and apply them to local landscapes in or out of protected areas (figure 1.1). But where do these conservation guidelines originate? Ideally, the people who will implement them will first consider the possible consequences of each reasonable choice and then select the alternative most likely to favor conservation goals while being acceptable to most local communities. By what means, though, can conservation professionals assess the likely consequences of each alternative? Can they simply follow their gut feelings? Sometimes—if and only if their insight into the landscape's natural

Figure 1.2.
A landscape managed for conservation
(Reserva Natural La Planada, Nariño
Department, Colombia).

Box 1.1. It Might Sound Good, but Will It Work for You?

In 1983, in the discussion section of a short paper in a widely read scientific journal, D. H. Janzen wrote the following sentence: "It is hard to avoid the conclusion that in some circumstances, it may be much better to surround a small patch of primary forest with species-poor vegetation of non-invasive species of low food value [e.g., grain fields, closely cropped pastures, cotton fields, sugar cane] than to surround it with an extensive area of secondary succession rich in plants and animals that will invade the pristine forest." Janzen clearly intended to provoke conservation biologists and managers into undertaking critical tests of this possibility in their own landscapes—note his words "in some circumstances." Instead of stimulating careful, site-specific evaluations, though, that sentence and a similar one in the article's summary have been lifted from context and cited uncritically by so many authors that they have achieved the status of a universal law for conservation professionals. The result has been an inordinate number of generalizations, and (rumor has it) even management policies all assuming that nearby second-growth habitats must be having a pernicious influence on conservation of protected areas throughout Latin America—without the idea being further tested.

Janzen's suggestion actually sounds eminently reasonable in terms of both common management objectives and natural history. Indeed, the suggestion may be the best strategy for conserving a given reserve's original vegetation under some circumstances. You'll see the paper cited at several other points in this book. Hold on, though! Before you too rush outside armed with machete, shovel, driptorch, and shotgun, eager to level all the nasty, species-rich second-growth vegetation and attendant animals within or nearby the area you manage and to replace these with innocuous soybeans, pasture grasses, or asphalt pavement, please reread the sentence and then read Janzen's entire paper. You'll find that the statement is a carefully couched speculation based on the author's incidental observation that plants of some second-growth species were colonizing one sunlit gap created by one fallen tree in one small remnant of one singular tropical dry forest in northwestern Costa Rica. Considering this, should you undertake a costly and drastic management plan for your reserve, which almost certainly differs in every conceivable way from Janzen's patch of Central American dry forest? In your particular landscape, might second-growth vegetation have positive as well as negative effects on the persistence of original vegetation and animals? Might the positive effects even far outweigh negative ones, if any? How might you find out, so as to come up with the best possible conservation guidelines?

In a similar vein, despite the appeal and biological reasonableness of the concept of "habitat corridors," some have questioned whether it is wise to rush to implement such corridors throughout temperate and tropical landscapes without having a better idea of their cost effectiveness, conservation effectiveness, and site appropriateness (e.g., Simberloff et al. 1992; Crome 1997; Schwartz 1999). It's likely that each case is unique. As Crome puts it, "Be suspicious of all but the most obvious generalities. Completely disbelieve the obvious ones."

history and social context is acute. Or should managers defer to "those who must know better" and base their guidelines on appealing, reasonable-sounding, and widely accepted ideas encountered in a published paper or heard at a conference? I certainly hope not (see box 1.1). Instead, might conservation professionals themselves evaluate the various alternatives in the very landscape where they would be applied (figure 1.2)? Yes, through thoughtfully designed and cautiously interpreted studies carried out firsthand (figure 1.1). How might such studies, on the ecological consequences of alternative management decisions, be designed well and interpreted cautiously? By using the approach of scientific inquiry.

Getting at Scientific Inquiry

Let's back up. What do *scientific inquiry* and *science* really mean? *Formal science* (or *basic science*) consists of two components that are linked by a dynamic process (figure 1.3). One component is the body of accumulated and continuously accumulating observations (data) that researchers generate with reference to the other component, the body of concepts that provide the current frame of reference. In turn, the body of concepts is constantly reevaluated and modified in light of the incoming data. The science process, or scientific inquiry as defined below, provides the means to cycle back and forth between concepts and data.

If science consists of a dynamic cycle, as in figure 1.3, is the isolated act of gathering data "science"? No. Does a long published list of observations (data) bereft of a conceptual context constitute science? No. Does sitting at one's desk and proposing a new theory make one a scientist? No. Does the use of sophisticated electronic instruments or complex statistical procedures justify applying the name "science" to any endeavor? No. Science requires that all four elements illustrated in figure 1.3 be present: the two boxes and the two arrows.

Combining the Two

In this book I'll stress *scientific inquiry,* the cycling process of figure 1.3, rather than dwelling overlong on details of the figure's two boxes. In the broad sense, scientific inquiry is *a means of asking and answering firsthand, as objectively and precisely as possible, a question about a small piece of one's*

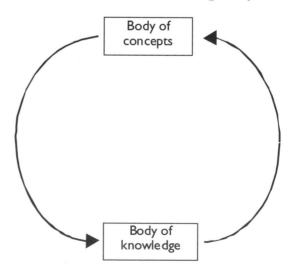

Figure 1.3.
Essential elements of formal science. Concepts and ever accumulating knowledge are related to one another through the process of scientific inquiry, as indicated by the arrows. The cycle may stand alone, as in "pure" or basic science, or may relate to applications such as technology or (in our case) conservation.

surroundings and then reflecting cautiously on the implications of the answer to the larger world. The quandary is that the concerns of conservation professionals and field ecologists often involve a fairly grand spatial scale on the one hand, and a fairly extensive time scale (the foreseeable future) on the other. In order to make the "correct" conservation decisions or the "correct" interpretations of ecological phenomena with absolute certainty, we'd have to be omniscient regarding that grand scale in time and space.

We aren't omniscient, though. We can't investigate simultaneously all possible individual organisms, populations, species, points in space, and landscapes of interest, nor can we evaluate the consequences of every possible variation of each feasible conservation guideline. We can only work in present time; we have only a fuzzy idea of those past events that might have caused present-day phenomena, and we certainly can't know for certain what the future portends. Thus, conservation scientists and others are restricted to working with "best guesses" based on the information available. That information comes from a *sample* restricted in space and time. We wish to *extrapolate* in as error-free a way as possible from that limited sample to the larger (and future) world where conservation guidelines will take effect or our conclusions might apply. Scientific inquiry guides us in (a) framing the question; (b) figuring out the most practical, meaningful, and objective way to obtain the sample so as to answer the question; (c) choosing the right tools to help get at the answer; (d) interpreting the results; and (e) extrapolating as cautiously as possible to the larger arena of conservation and management decisions.[3]

This book is meant to guide you, the reader, in the practical use of scientific inquiry as a tool for conservation in protected areas and in the "semi-natural matrix" (Brown, Curtin, and Braithwarte 2001), or for studies in basic field ecology, wildlife biology, and related fields. Along the way, the text will address a number of specific concepts, approaches, useful quantitative tools, and concerns or warnings. This will be done, I trust, in a commonsense rather than a technical manner. For example, frequently I'll raise concerns about approaches, techniques, and labels that are currently fashionable in biodiversity conservation, not in order to discredit them forever but so that you will think twice before rushing out to apply them uncritically to your particular landscape. I'll also point out that basic science has a great deal to offer the practical realm by helping to provide the conceptual frameworks that generate conservation concerns and important management questions (Poiani et al. 2000; The Nature Conservancy 2000). Likewise, teamwork between trained ecologists and managers (who may be one and the same person), plus other professionals such as sociologists, can streamline the sequence illustrated in figure 1.1. Nevertheless, as you'll see in chapters 2 and 10, the tool of scientific inquiry is by no means the exclusive province of professionals (Cooperrider 1996), nor does its effective use depend on an intimate familiarity with the two boxes of figure 1.3.

How to Use This Book

Speaking of questions, if you're an ecologist or conservation biologist, do you believe that you're doing Real Science, and do you feel comfortable with that? If you're another sort of conservation professional, do you see any value whatsoever in applying scientific inquiry to conservation concerns? Do you feel at ease when you hear the word *science* or *the scientific method*? Whatever your answers, chapter 2 will encourage you to scrutinize them.

Whatever the theme that concerns you in the field, are you confident in your ability to phrase the crucial question in such a way that it can be answered directly and clearly firsthand and the answer then extended to the larger scale? In chapter 3, you'll get some practice.

Can you design the field study that will best answer your question, matching the scope of data collection to the scope of that question? Can you make the best possible compromise between a study that's perfectly designed but excruciatingly difficult to complete, and one that's easy to complete but so flimsy or biased that the answer does you no good? If not, chapter 4 might help. Even if you're delegating the actual study design and data collection to others, in a sense chapter 4 plus chapter 6 make up the heart of this book.

Have you ever encountered (a) confidence intervals for an estimate of something like the mean length of caimans in a lake, (b) a statistical test or a "*P*-value," (c) the word *significant* or the word *sample*? Are you perfectly satisfied with the particular set of results you obtained, period, or do you wish to use them to draw conclusions about, or apply decisions to, a larger universe in space and time? Are you aware that if you do calculate confidence limits on estimates or perform a statistical test, you always run the risk of drawing the wrong conclusion? Have you thought about the consequences to conservation policy and management decisions of unknowingly making such an error? Do you know any practical means to lower the risk of making such an error? Do you recognize the vast, and critical, difference between statistical significance and biological significance? Chapter 5 addresses these and many related doubts.

Please don't be put off by the subject matter of chapter 5—statistical inference—or the sudden appearance there of some mathematical equations. First, statistical inference is, or should be, based simply on common sense. Second, in chapter 5 you'll start with the most basic yet biologically important questions—how to present average values and the amount of variation among the data you've collected—and then build, step by step, your capacity in the logical philosophy and real-life application of statistical inference. Along the way you'll come to realize that statistical inference is more often misused and misinterpreted (again with potentially drastic consequences) than not, and that it may even be inappropriate to some questions in conservation, management, and basic ecology. If you're a student, you'll undoubtedly relish the thought of trying to convince your professors of this. Still, if you don't believe the assertion that the design of field studies (chapter 4) and statistical inference (chapter 5) are based on common sense yet are critically important to management decisions, skip to appendix D. You won't find a single mathematical equation there, but by the time you finish reading and acting, you may have grasped the most important fundamentals of both.

Speaking of the most important fundamentals of all, do you take natural history into account when choosing conservation guidelines or designing and interpreting studies? Are you adept at taking the point of view of the animals, plants, or landscapes you're trying to conserve or study, rather than unconsciously imposing your own? Again, the math- and technique-free chapter 6 is half of the heart of the book, so don't halfheartedly stick with chapter 4 only (or chapter 6 only, for that matter).

Do you work exclusively in protected areas or exclusively in the altered landscape outside of protected areas? Have you recognized, and have you taken into account, the ways in which those two landscapes—indeed, any two distinct habitats or landscapes—interact with each other? Chapter 7 presents some related points for you to ponder.

Do you realize that if you were to take into account everyone's point of view on your landscape's "ecological integrity," you'd have to monitor everything from bacteria to spectacled bears or jaguars? To simplify, and to attract public attention, do you base your conservation strategy on the concept of flagship species? Umbrella species? Keystone species? Indicator species? Have you clearly distinguished between species that are "indicators" and species that are "important," and have you thought of the consequences of basing conservation decisions on the latter rather than the former? Have you chosen the best possible species as indicators? Have you considered monitoring ecological processes, rather than species, to indicate "ecological integrity"? Chapter 8 will deal with these practical issues.

Before reading the title of this book, had you ever used, or heard, the word *biodiversity?* Like many others, did you equate biodiversity with species diversity? And, have you ever elegantly quantified species diversity with a well-known numerical index? Is the symbol H' or the name Shannon-Weaver index (or just plain Shannon index) familiar? Have you considered the basis, biological assumptions, and checkered histories of species diversity indices, or whether an index is even an appropriate way for expressing biodiversity? Are you aware of more informative alternatives? Chapter 9 will deal with these and other manifestations of "biodiversity."

Does the conservation you practice, or your field research, ever involve people who are neither conservation professionals nor basic ecologists? Have you considered involving a much broader audience in the philosophy of scientific inquiry? Schoolchildren? Park guards? Visitors? Community members? For some ideas, go to chapter 10. The last chapter won't make sense, though, if you've skipped the rest of the book. So, to start this off, let's return to the practical aspects of scientific inquiry. How does this investigative tool work, and how can it be made accessible to people not specifically trained in Real Science?

CHAPTER 2

The Inquiry Process

If science is to help in biological conservation, it must be a much more inclusive and widespread sort of science than we know now.

—Allen Y. Cooperrider (1996)

Any scientific inquiry whatsoever begins with a question about the features of one's surroundings. Depending on the interests of the investigator, the scale of those features might be that of subatomic particles, genes, whole organisms, species, entire landscapes, continents, our solar system, or the galaxies making up the universe. The questions that many ecologists and conservation professionals have, and the concerns that generate the questions, tend to involve the spatial scale of landscapes even if the focus is on a single species. In particular, conservation concerns might arise from a number of sources. Most have to do with how the protected area, its surroundings, or species of concern might be affected by different events, threats, or, of course, possible management guidelines—the last including alternative ways to counter a given threat (see figure 1.1). For example, your concern might involve the amount of firewood that can be extracted from the reserve without seriously compromising plant regeneration and soil quality, or the choice among alternative management guidelines that will lead to the most rapid recovery of a watershed from illicit gold mining activities. The key to enlightened conservation of protected areas or altered landscapes, just as to enlightened research in wildlife biology or forest ecology, is knowing how to frame questions. I urge you to go through the exercise of box 2.1 before reading any further.

Box 2.1. Practice with Observing the Landscape and Framing Questions

Grab a notebook and pencil. Go outdoors and find a miniature landscape. This could be a flower box, a patch of weeds bordering the street or trail, the visitors' parking lot at a protected area, moss-covered stones, an abandoned field, a school yard, a cow pasture, the ground beneath a forest canopy, or the trunk of a large tree. Select and mark off a small parcel, of about 50×50 cm, that displays a fair amount of "patchiness" (heterogeneity) within its borders. First, carefully examine the landscape you've just demarcated. Then spend about five or ten minutes sketching a crude map of the major "ecological elements" the landscape contains—for example, different types or forms or patches of plants, bare soil, leaf litter, insects and spiders, stones, fallen twigs, cracks in the cement, pieces of trash, patches of sun and shade, crevices in or lichens on a tree trunk. Then, spend ten minutes or so thinking up and writing down at least five *questions* (more if possible) that spring to your mind regarding what you've noted within the parcel's boundaries. No restrictions apply to the subject matter or format of the questions. Feel free to poke at things with your fingers or a stick. Most important, don't hesitate to write down any question that occurs to you. Rule Number 1 is, *there is no such thing as a stupid question*. Some questions might lead more easily than others to firsthand inquiry (the theme of chapter 3), but all questions are valid as such. Once you feel at ease posing questions about what you see, you've mastered the most critical phase of scientific inquiry.

The Formal Scientific Method: Too Academic?

Now that you're experienced at generating questions at the level of landscapes, albeit landscapes in miniature, let's discuss scientific inquiry as it might proceed from such questions. By any definition, scientific inquiry involves progressing through a series of logically related steps that eventually allow one to provisionally answer, or to revise, the original question as objectively as possible. Formal science and trained scientists, in theory at least, employ a detailed scheme called the *scientific method* or the *hypothetico-deductive method* (figure 2.1).[1]

In the formal scientific method, the question loses the punctuation and becomes a declarative statement, the *prediction*—but only after going through the three stages that make up the top line of figure 2.1. First, either on its own or stimulated by a firsthand observation about the scientist's immediate surroundings, a general concept (theory) or working frame of reference (paradigm) suggests to her that a particular relationship, pattern, or effect might occur in the universe at large, including but by no means restricted to those surroundings. This proposal is then formalized as the scientific hypothesis, or more accurately the scientific alternative hypothesis (scientific H_A). Naturally, the investigator recognizes that this possibility is only one of two, the other being that no such relationship, pattern, or effect really exists (the scientific null hypothesis H_0). Please note that scientific hypotheses are entirely distinct from the statistical null and alternative hypotheses that are discussed in chapter 5.

The scientist cannot, of course, evaluate the two scientific hypotheses or the theory that generated them under all possible conditions of space and time where they might apply. She has only the immediate surroundings, and the present time, for such a test. Therefore, the third and final step of

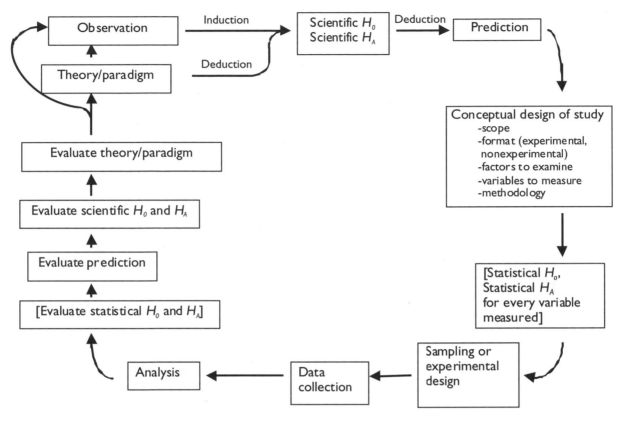

Figure 2.1.
The formal scientific method. Here, scientific inquiry begins with the two boxes at the upper left:
the theory or paradigm and the observation.

the questioning phase is to derive (deduce) a prediction that narrows the scope of the scientific (alternative) hypothesis to those immediate surroundings accessible to an investigation. The prediction specifies, "If the scientific (alternative) hypothesis is always true under Conditions X, Y, and Z, and my immediate surroundings present Conditions X, Y, and Z, then such and such will occur here and now." Here's a rather obvious example of this logic:

- PARADIGM: Vegetation interacts with the water cycle. Or,

- THEORY: Vegetation cover affects the rate of evaporation from the soil surface.

- OBSERVATION: I work in a forested reserve, and selected parcels will be clear-cut as part of the management plan.

- SCIENTIFIC ALTERNATIVE HYPOTHESIS: Clear-cutting changes the rate of evaporation from the soil surface.

- SCIENTIFIC NULL HYPOTHESIS: Clear-cutting does not change the rate of evaporation from the soil surface.

- PREDICTION: *If* clear-cutting changes the rate of evaporation from the soil surface anywhere and anytime, *and if* selected parcels are clear-cut in the forest reserve where I work, *and if* I measure evaporation rates from the soil surface before and after clear-cutting, *then* I will record a change in those rates.

After specifying the prediction, the investigator moves down the right-hand side of figure 2.1 and then across the bottom toward the left, proceeding cautiously through the steps of study design, data collection, and data analysis. The results of the data analysis then allow her to clamber back up the left-hand side of figure 2.1, first evaluating the statistical hypotheses—if using statistical inference (chapter 5)—and then scrutinizing each of the first three steps, those on the top line, in reverse order. That is, if the data she's collected and the statistical analysis of those uphold the prediction, this result provides one more bit of support for the all-inclusive scientific (alternative) hypothesis, which in turn provides one more bit of support for the more general concept (theory) or the even more general frame of reference (paradigm). On the other hand, in formal science a convincing failure to support the prediction *in even a single instance* means that the scientist cannot reject the scientific null hypothesis; that in consequence the scientific alternative hypothesis is not supported in this instance and therefore cannot be universally true; and that the body of theory itself, even the paradigm, must be reevaluated and modified, leading in turn to new passes through the cycle of figure 2.1.

Scrupulously followed, the formal scientific method is an extraordinarily powerful means for advancing basic science in the manner of figure 1.3. Is the formal scientific method always the best framework for scientific inquiry, as defined previously? Perhaps not.

First of all, the formal scientific method is inappropriate to the goals of conservation initiatives, such as protected area management, or to goals of most field research in ecology and related fields (Crome 1997; Johnson 1999). The formal scientific method emphasizes scientific investigation as a direct means of evaluating the upper left-hand corner of figure 2.1 or the top box of figure 1.3— the body of concepts (theory)—rather than focusing on the particular prediction or specific question that has to do with one's surroundings. That is, if she faithfully follows the formal framework of figure 2.1, the scientific investigator exploits predictions and their tests (research projects) simply as mechanisms for evaluating the much grander "yin and yang" of the two universal scientific hypotheses. In this manner she then gains fame (though rarely fortune) either by providing one more bit of support for the universal truth of the scientific (alternative) hypothesis and the general concepts currently in vogue, or else by soundly thrashing the alternative scientific hypothesis and forcing a severe modification of the theory. In this scenario, the forest tract whose parcels are clear-cut is of little interest in itself. Rather, it's a single "trial" for evaluating the universal scientific (alternative) hypothesis that vegetation cover affects the rate of evaporation from the soil surface anywhere and everywhere.

If you're the conservation researcher or protected area manager, are you more interested in the universal validity of the scientific hypothesis regarding vegetation cover and evaporation rate, or in the particular consequences of logging within the particular reserve where you work? If you're a field ecologist examining the relationship between food scarcity and interspecific competition in hummingbirds, are you more interested in the universal validity of the statement that food scarcity and interspecific competition are related for any organism in any context, or in what's happening with the hummingbirds within the particular habitat where you work? Let's consider another conservation professional working in a Latin American protected area. This person is worried not about the set of all protected areas worldwide, but about the one for which he is responsible. It doesn't matter to management of this particular protected area whether the most effective means of restoring a watershed degraded by gold mining here is also the most effective means in Siberia, Zimbabwe, Australia, or even on the other side of the mountain range. The urgent question he is asking concerns the immediate surroundings only, where the answer in turn will guide policies pertaining specifically to those surroundings (see figure 1.1). The first two steps illustrated in figure 2.1—the general concepts along with the observation, if any, that catalyzes the process and the universal scientific hypotheses—should simply serve as a convenient means of getting to, and framing, that urgent question whose scope is restricted in time and space. And rarely, if ever, will the conservation professional benefit by turning a clearly presented, open-ended question (see chapter 3) into a rigid, formalized, declarative prediction.

The Inquiry Cycle: Too Simple?

Second, let's face it, the formal scientific method as it is usually presented (for example, in figure 2.1) intimidates most people, including many trained in Real Science. Even if its complex terminology and apparent philosophical rigidity don't frighten people off, how many have simply memorized (or used blindly) the scientific method without any real understanding? Please squint suspiciously at figure 2.1. All those jargon-filled boxes, the fundamental concepts of the formal scientific method, can be collapsed into four basic, logical steps: the "inquiry cycle" (figure 2.2). Let's rephrase the preceding section in commonsense language and focus it on answering our specific questions rather than on evaluating universal concepts.

As always, inquiry begins with an observation about one's surroundings. The observation never stands alone, though. Consciously or unconsciously, the observer always places it in the context of a broader concept or concern. That context need not be formal at all. For example, you observe a two-legged being of about your size holding a slender, long object made of wood and steel, approaching several large feathered things sitting in a tree. Based on previous experience, you place these observations in a broader context: a hunter from the local community is about to shoot a chachalaca (charata), a large edible bird. The observations placed in context stimulate you to *construct the question*—just what you did in the exercise of box 2.1. In this case, the question that leaps to your mind might be, "How does the abundance of chachalacas in the forest change with respect to distance from communities?"[2] In essence, the entire top line of figure 2.1 collapses into the first step of figure 2.2: coming up with the question.

Next, as the investigator you take *action* to answer the question, by designing and carrying out a study at the scale to which the question refers. This step, which covers the right-hand side and bottom line of the cycle in figure 2.1, results in a set of findings or data.

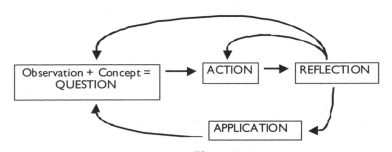

Figure 2.2.
A different scientific method: the simple "inquiry cycle."

Once the action step has provided a specific answer—the observed results—to your particular question, the crucial third step of *reflection* takes place. In essence, this step covers the entire left-hand side of figure 2.1 and more. Reflection takes several forms, as the leftward pointing arrows in figure 2.2 indicate. What do your findings imply, with respect to the original question? Has your initial reasoning been supported, or not? Could other phenomena, not originally considered in the question, have produced the same results? Should the question have been oriented differently? What are the conclusions you're willing to draw? Just how far, with respect to other locations, times, or conditions, can you justify extending your conclusions and speculations? How confident do you feel about basing real-life decisions on those conclusions and speculations? Did events during the action step, or the results themselves, bring up new, intriguing, or urgent questions, each capable of initiating its own cycle of inquiry? Finally, you should closely reexamine the action step itself, especially the way you designed it and carried it out. Was the study fully adequate to answer the original question? Might the results have been influenced by biases in the methodology or sampling? Might unforeseen complications of natural history, behaviors of animals, features of plants, subtle ecological interactions, or weather events have influenced the particular results such that the question as originally posed hasn't yet been answered directly?

The reflection step is as crucial to questions in biodiversity conservation as it is to those in any branch of basic science. Far too few inquiries, whether carried out by laboratory scientists or conservation professionals, include adequate reflection. Results are summarized and "written in stone," conclusions are drawn, without considering further their validity and significance with regard to the original question. Why is this a problem? Because, in the particular case of protected area management or other conservation endeavors, the inquiry cycle includes a fourth step: *application*. The results and interpretation of a study strictly circumscribed in time and space will guide practical decisions affecting larger scales and future times. These decisions may have far-reaching effects on the destiny of the landscape in question. Therefore, the reflection over "how far can I extend my speculations and conclusions" takes on tremendous importance. Furthermore, the reflection process must continue once the application takes effect, once conservation guidelines that are based on careful scientific inquiry are implemented. That is, the effect (and effectiveness) of those guidelines should constantly be monitored and reevaluated. Some effects may be unexpected, giving rise to further concerns and questions that in turn catalyze new cycles of the entire four-step process illustrated in figure 2.2.[3]

Is the "inquiry cycle" diagrammed in figure 2.2 sufficient for accomplishing basic science (figure 1.3)? You'd certainly find disagreement on the answer to that question among philosophers of science: many would find figure 2.2 too simplistic. As a professional researcher, though, consciously or unconsciously I've based my own investigations on it; as a teacher, I've exploited the inquiry cycle with various groups ranging from elementary school teachers to postgraduate students in ecology and conservation professionals. If you're dealing with questions not in conservation and management but in field ecology or other basic studies, you simply need to change the word "application" in the fourth step to "wider universe" (Feinsinger, Margutti, and Oviedo 1997). That is, just how far can you generalize from your results to phenomena that might happen, or have happened, in other places or at other times? Later in the book I'll often refer back to this four-step approach, for example, when discussing means for bringing local communities into the inquiry process (chapter 10). Most people use some form of the inquiry cycle, often labeled "common sense," in their daily lives. Nevertheless, you're probably a bit skeptical, muttering to yourself, "Well, yes, the scheme in figure 2.1 is no use to me because it's too academic, but isn't the one in figure 2.2 also useless because it's oversimplified?" Before we address those doubts, let's consider one case of this four-step scientific process (figure 2.2) as practiced in the field.

Do Campesinos Practice Scientific Inquiry?

A farmer (campesino) living in the wet highlands of northwestern Ecuador, or nearly any other landscape with small-scale agriculture, must make "management decisions" almost daily. For example, he must decide which cash crop to plant in a one-hectare parcel of recently cleared land on a certain hillside (figure 2.3). Because of the time and energy required to plant and cultivate the parcel, he wants to ensure beforehand that the crop he chooses will yield under the problematical conditions of nearly constant cloud cover, cool temperatures, and high humidity that characterize the parcel. Having other parcels to tend, he is in no particular hurry and would rather wait a year than rush into making a wrong decision regarding the new clearing. He decides to make a limited trial of two crops: a tuber crop, melloco (*Ullucus tuberosus*), and a legume crop, chocho (*Lupinus mutabilis*). So, after preparing the plot, he plants a few melloco tubers or chocho seeds in each of several spots scattered throughout the parcel—well aware that there's variation in soil and moisture conditions over the parcel itself, and that he wouldn't learn anything useful by confining the trial to one corner only or by planting the chocho in one corner and the melloco in another (see chapter 4). During the growing season he occasionally visits the site and checks on the growth and health of each crop. At harvest time, he discovers that a fungus has attacked most of the chocho plants and pod production is near zero, whereas the melloco plants seem healthy and have produced reasonable numbers of tubers.

Clearly, then, *under the conditions of his trial*, the melloco makes the better choice. Before running out to purchase enough melloco tubers to plant the entire hectare, though, the campesino ponders a bit. Have weather conditions of the past year been typical, or was the year unusually wet, perhaps favoring the growth of the fungus and causing an atypical failure of the chocho crop? Is it reasonable to expect next year's conditions to resemble those of the past year, or not? Also, he has a different one-hectare parcel at a lower elevation, with exposure, moisture conditions, and slope quite different from the parcel he has just evaluated. Do the results from the first trial permit him

Figure 2.3.
An agricultural mosaic in the Cordillera de Toisán in highland Ecuador (Imbabura Province). Some parcels are used for subsistence crops or livestock pasture, others for cash crops.

to assume that melloco will be the superior crop for this rather different parcel as well? After some thought, the campesino decides to take the risk and plant melloco in the first 1-ha parcel where he carried out the trials, despite uncertainty about next year's weather. He also decides, though, not to risk planting melloco in the second, quite different parcel until he has had the chance to undertake analogous trials there, perhaps involving other potential crop species.

Compare the sequence of steps (observation and doubt, trial, reflection, and decision) through which the campesino progressed, from the starting point of wondering which crop might do better through the end point of planting the selected crop, with the inquiry cycle illustrated in figure 2.2. Has he practiced scientific inquiry? Yes, absolutely. Has he undergone formal training in science? Almost certainly not. He has three neighbors, who base their own decisions about which crops to plant where on, respectively: (1) the advice of the newly arrived agricultural extensionist, (2) the opinions of a great-uncle whose farm is in another watershed, and (3) the advice of an almanac. Are the neighbors also engaging in scientific inquiry, or not?

The "Management Cycle" and the "Field Research Cycle"

Now, let's develop two compromise frameworks for scientific inquiry, somewhere between the academic formality of figure 2.1 and the simplicity of figure 2.2. First, what if you're a conservation professional? You need a scheme for arriving at the "best guess" regarding alternative possibilities for guidelines. Figure 2.4 presents one such scheme.

Here, the inquiry process begins with a particular concern. This might arise from any one of a number of sources. For example, you might work in a protected area and observe directly a current or potential threat to the reserve's integrity, a "stress" in other words (The Nature Conservancy 2000). Or your concern might arise from prior knowledge (your own or someone else's) of what *might* be a problem. For example, maybe you've just read the article by Janzen (1983)—see box 1.1—and now wonder if the abandoned pastures, laden with weedy pioneer shrubs, that border on the reserve's primary forest might be overwhelming the forest's soil seed bank and altering its internal vegetation dynamics, in comparison with the well-maintained, weed-free cattle pastures or cornfields bordering other parts of the reserve. Or your concern might simply arise from commonsense reasoning. For example, given that selective logging occurs in the reserve's perimeter and that a choice exists between oxen and mechanical skidders for extracting the logs, it makes sense to wonder if one or the other choice will wreak less damage. Or the concern may arise from local community members (chapter 10). Perhaps they wonder just how much exploitation a critical population of medicinal plants can tolerate without dying out. Finally, the concern might originate far from the protected area itself: for example, the central ministry has just sent you a directive to evaluate a certain management policy or a certain type of perceived threat.

The management concern generates that all important step: framing a question that will be answerable through firsthand inquiry (figure 2.4). In chapter 3 we'll discuss the ins and outs of framing questions. A well-framed question leads readily to the conceptual design of the inquiry proper, which chapter 4 will explore at length: which data to record, where, when, how, and by whom? What's the scope of your question? How can the inquiry be designed so that you can cautiously apply its results over that entire scope? Does the question require an experiment or not? What are the factors or relationships you'll evaluate? What particular sorts of data will you collect? What methods will you use to collect them? Finally, will inferential statistics be useful? If so, before proceeding

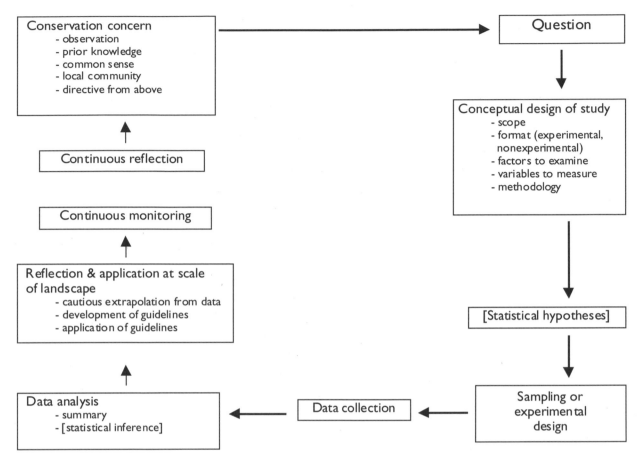

Figure 2.4.
The "management cycle": The inquiry cycle of figure 2.2 expanded for scientific inquiry
applied specifically to conservation concerns and other questions that will lead to guidelines
or management decisions. The process begins at the upper left.

further you must set up formal statistical hypotheses for each sort of data you'll collect, as we'll discuss in chapter 5. Whether or not you'll use inferential statistics, how many samples should you take or experimental trials should you run? Chapter 5 and appendix B address this crucial and sometimes overlooked decision.

At last you collect the data. After analyzing, summarizing, and (if required) applying statistical inference to the data (figure 2.4), you must sit back and reflect, just as in the third step of figure 2.2. What do the results really mean? Again, given that the data represent the here and now, whereas conservation decisions will affect the there and later, just how far can you apply the conclusions you've drawn? Chapter 6 will consider this crucial question from the viewpoint of natural history. Finally, following reflection, you and others concerned use your inquiry as one piece of information for proposing conservation guidelines and putting them into practice. That's not all, though. As before, you must continue to assess effects of the guidelines by means of a rigorous monitoring program. Eternal monitoring is the price of conservation. Eternal vigilance is the price of monitoring, though, and rather than simply accumulating great gobs of monitoring data, you must continually evaluate and reflect upon those data as they accumulate. Again, unforeseen side effects of the chosen guidelines may generate new concerns, which in turn should initiate new cycles of inquiry.[4] Please note again the critical difference between the "management cycle" (figure 2.4) and the formal scientific method (figure 2.1). Here, your inquiry revolves around your own landscape and the events therein. You're not out to support or refute the effectiveness of particular guidelines the world over.

Likewise, if you're a student or professional in field ecology, wildlife biology, or a related field, would you be perfectly comfortable in considering your lengthy and difficult research project as just one tiny bit of information used to evaluate a universal scientific hypothesis without any other intrinsic value? If your final results fail to support the prediction that interspecific competition among hummingbirds and food abundance are related at your site, are you prepared to trumpet to the world, "The scientific (alternative) hypothesis, of a universal relation between interspecific competition and abundance of resources, was not supported. Therefore, we must abandon that hypothesis forever and reevaluate the theory and paradigm of interspecific competition"? I doubt that you'll answer yes to either question. Sorry, but if you're an ecologist or other field biologist, you're not doing Real Science in the formal sense of figure 2.1. Instead, you recognize the complexity of field ecology and natural history and recognize that every situation is likely to be unique because of the myriad other factors influencing what goes on here and now (Crome 1997). You're most interested in a particular question regarding these hummingbirds in this landscape and at this time, although in the discussion section of your thesis or paper you'll be happy to speculate on how your results might apply to events elsewhere.

Actually, though, you *are* engaging in Real Scientific inquiry—not that of figure 2.1 but rather that of figure 2.5. You're following the "field research cycle." Observing hummingbirds sparring with one another over a flower patch and knowing that the number of flowers is likely to vary over the year, you've thought about the concept of interspecific competition, and an idea for a scientific study begins to form. That general idea—and please don't confuse the issue by labeling it "hypothesis"—leads to a specific question (see chapter 3). You'll answer the question by designing the particulars of the study, collecting data, and reflecting on all aspects of the results—including whether they fit with general theory and with results of other studies on the theme. The major difference between the cycle you follow as a field ecologist (figure 2.5) and that which you follow as a conservation professional (figure 2.4) is that this time you're not going to apply your results to management decisions on hummingbirds or flowers, or to monitoring the consequences of such decisions.

Any of the four methods presented has potential for guiding scientific inquiry. Take your pick. Perhaps the goals of figure 2.5 are most appropriate to your interests. Perhaps the applied goals of figure 2.2 or figure 2.4 are most relevant to your conservation concerns. Perhaps you're a physicist and figure 2.1 suits you just fine. Whichever the scheme used, though, the step of framing the question is critical, simply because some ways of framing questions lead to much more useful reflections and applications than others do. In the next chapter we'll focus exclusively on that step.

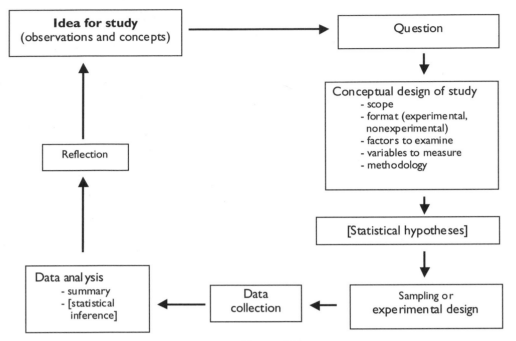

Figure 2.5.
The "field research cycle": The inquiry cycle of figure 2.2 expanded, or scientific inquiry applied to questions in basic field ecology and related approaches.

CHAPTER 3

So, What's the Question?

Asking the right questions is as important as answering them.

—Allen Y. Cooperrider (1996)

As figure 2.4 points out, conservation concerns or other sources (for example, the exercise in box 2.1) may spark numerous questions. Again, in one sense all these questions are valid. In another sense, though, some may be more valid than others. That is, in real life some questions are asked with the intent that they will generate clear answers that in turn will lead to effective conservation guidelines or solid scientific conclusions.

Please understand that this chapter and the book as a whole remain quite silent on the issue of which questions are the most important ones for conservation researchers—or others—to ask. You aren't reading a guide to conservation policy or to the hot topics in field ecology. It would be a serious mistake to pretend to instruct conservation professionals and field biologists on which issues or strategies they should investigate. Likewise, the examples listed later in this chapter are meant only to illustrate some questions that comply with the four criteria discussed below, not to serve as exemplars of the most burning issues to address. Now let's discuss those criteria: how might you frame your question so that it leads most readily to the complete inquiry cycle (figure 2.2), management cycle (figure 2.4), or field research cycle (figure 2.5)?

Framing Answerable Questions

A question should be phrased so that it's directly answerable by collecting data within a reasonable time frame. Questions such as "which?" "how many?" "where?" "when?" "what's the relation between?" "what are the immediate consequences?" and "what are the differences?" can often be answered directly through careful application of the scheme in figure 2.2, 2.4, or 2.5. They direct you to investigate patterns or events in the present day.

How many times during the exercise in box 2.1, though, did you frame a question around the word *why*? In contrast to others, "why" questions are at the heart of the reflection step (figure 2.2) but inappropriate for construction. You'll see this if you try to initiate a scientific inquiry with a why question. You'll be stuck. A why question directs you to detail the unknown events in the past that led to what you observe today. Unless you own a time machine, you can't do that. The why questions pop up when you're reflecting on your results, though. You don't answer them directly; rather, you propose possible explanations for your observations, you speculate, you consider alternative possibilities. Most important, why questions prod you to transform them into new, answerable questions that initiate new cycles of inquiry.

For example, the question "*Are* more seedlings of weedy pioneer plant species invading the reserve where it is bordered by second-growth vegetation than where it is bordered by cattle pasture?" is answerable through direct investigation, today. The question "*Why* are there more seedlings of weedy plant species in the first situation than in the second?" is what springs up if you've answered yes to the first question. The why question, though, can't be answered directly today, because it involves events in the past that may have influenced the arrival, survival, and germination of the seeds that produced the seedlings you see today. One possibility is that proposed by Janzen (1983), as described in box 1.1. That is, it's possible that in the recent past the second growth and the fruiting plants it contains had attracted fruit-eating birds and mammals, which, when returning to the reserve, defecated large numbers of those plants' seeds. As a natural historian, though, you realize that any number of other past events could also account for a present-day difference in seedling densities between different sites. These possibilities include differences in land-use histories, in rates of pathogen or fungal attack on seeds that lie buried in the soil, in seed predation by ants and rodents, in the rate at which seedlings themselves were attacked before you happened along, in exposure to wind, in other microclimatic features, and in the soil properties themselves—as well as pure chance. To answer the why question directly, you'd need to take the impossible step of examining all of these possibilities and many more besides.

Nevertheless, *reflecting* on why you might have obtained those particular results might lead you to consider seriously one of the possible explanations, frame an answerable question about what's happening right now with seeds or seedlings, and answer that new question with a firsthand inquiry. For example, let's say that you've observed that most of the invasive plant species do indeed have fleshy fruits and animal-dispersed seeds. Reflecting on this, you might propose that Janzen's (1983) scenario is indeed the most likely explanation for the pattern you observed previously, and the most urgent to evaluate before deciding on a management plan. Thus, the why question that surfaced by reflection during your first inquiry cycle (concerning patterns among seedlings) has now generated a second cycle. The second cycle might begin, for example, with the question "Do animal feces containing seeds of the invasive species arrive more frequently in forest bordered by second-growth vegetation than in forest bordered by cattle pasture?"

Avoid the temptation to begin a scientific inquiry with a question that really belongs in the reflection phase. For example, the questions "How might we manage the reserve so as best to maintain its biodiversity over the next century?" and "How can we design the education center so that it best encourages visitors to conserve their own surroundings upon returning home?" are urgent, compelling questions. They're questions for reflection, though, best answered by sitting around a big table and reaching a group consensus.

Framing Comparative Questions

A question should be a comparative one. The comparison should be that implied by the management concern (figure 2.4, upper left) or by whatever other concept generated the question (figure 2.5). A comparative question requires that you focus on that concern or general concept and allows you to address it during the reflection afterwards. By contrast, a noncomparative question is often a dead end, providing no basis for reflection or for proceeding further.

Let's imagine that your concern involves habitat corridors (see Beier and Noss 1998). The question "How many individuals and species of mammalian carnivores use this habitat corridor?" complies with the first criterion presented above: it can be answered by collecting data firsthand. Nevertheless, those data lead absolutely nowhere in terms of developing conservation guidelines. In contrast, the questions "Do more individuals and species of mammalian carnivores use a corridor of second-growth vegetation or one of primary forest?" "Do more mammalian carnivores cross an agricultural landscape if it has a habitat corridor than if not?" and "Is the corridor used more frequently at some seasons than at others?" all involve comparisons between two or more sets of conditions. Reflecting on the answers to these questions might lead to broad-scope guidelines regarding habitat corridors.

Of course, data gathered objectively in response even to a noncomparative question are valuable in their own right. Occasionally, it's impossible to muster the time and resources to examine the two or more sites, times, sets of conditions, or experimental treatments required by a comparative question. Nevertheless, by amplifying just a little bit the scope of a question that was initially noncomparative, the conservation professional can often greatly enhance the usefulness of results.

Consider a second example, a recently designated protected area that presently consists of a number of fragments of primary forest scattered in a sea of second-growth scrub. The overriding conservation concern is to maintain a high diversity of forest primate species. The reserve's conservation scientist looks at one particular forest fragment and asks, "How many primate species live in this forest patch, and of those, which are rare and endangered?" The question is valid, but will the results be of any use other than for describing the status of the particular fragment? If she changes the question to "Is there a difference in the number of rare and endangered primate species living in this fragment and in that other fragment of about the same size?" the investigator now has a comparative question, but it doesn't go much further than the first one. The comparison doesn't address the concern whatsoever. It involves no broad context, no general factor, no basis for making management decisions about the numerous other forest fragments scattered throughout the new reserve. Really, the second question is simply the first, noncomparative question squared.

If the investigator instead asks, "Is there a difference in the number of rare and endangered primate species living in this *small* fragment and that *large* fragment?" she's bringing in a much broader context (the concept that patches of different sizes may support different numbers, and identities, of resident species). She's also bringing in a possible key to management decisions: if indeed she finds a difference, then habitat fragments might be manipulated in particular ways to produce particular primate assemblages.

Of course, upon reflection, the investigator in this case realizes that examining only two forest fragments is not sufficient to generate guidelines. An observed difference in the two primate assemblages might have arisen from unique features of the two patches, features that have nothing to do with their distinct sizes (see chapter 4). Therefore, before proposing and applying management guidelines, she decides to survey a much greater number of habitat fragments that differ primarily with respect to the factor of size. So, her comparative question has now become "Do forest fragments of different sizes support different numbers of rare and endangered primate species?"

You may be chuckling over the naïveté of the investigator, saying to yourself, "What a stupid question! Everyone knows that larger forest patches have more species!" Please note that this real-life question is far from being stupid, though, and remember Crome's (1997) quote from box 1.1: "Be suspicious of all but the most obvious generalities. Completely disbelieve the obvious ones." In her particular landscape, the conservation scientist can't automatically assume that large forest patches will hold more endangered primate species than small patches just because several quite convincing theoretical papers tell her that such should be the case or just because results of some field studies undertaken in other landscapes happen to support that "conventional wisdom." It's quite possible that in her particular case the regenerating, resource-rich, second-growth scrub, or the border between forest and scrub, supports such a large number of primate species that small or intermediate-size fragments embedded in that matrix actually support the richest faunas, whereas large tracts of rather homogeneous forest support primate assemblages that are less diverse. The investigator won't know, and can't mount management guidelines, until she has dealt with the question firsthand by designing a study within her own landscape. Note that the criterion of "comparativeness" is just as critical to questions in basic field ecology as to conservation questions.

Framing Alluring Questions

The question should be an alluring one. That is, it should involve neither an answer that's already obvious nor an action step (figure 2.2) so tedious or time consuming that the data will be irrelevant by the time they're finally compiled. For example, the question "Which supports more native frog species, 700 hectares of marsh and swamp, or the 700-hectare paved parking lot for visitors?" is both answerable and comparative, thus complying with the criteria above. It doesn't merit a firsthand inquiry, though, as the comparison is nonsensical.

On the other hand, a question could comply well with all the foregoing criteria yet fail the test of allure, simply by requiring an inordinately long time frame or an inordinately complex plan for data collection. For example, consider the question "Which management tactic will result in the greatest diversity of canopy trees in next generation's forest, hand-planting seedlings of a selected diversity of primary forest species or allowing natural regeneration to take place?" The question could have extraordinary import to the long-term conservation of protected areas. In theory a definitive answer would settle the choice between two very different management approaches. Nevertheless, obtaining that answer through a well-designed firsthand inquiry might require several centuries and be of little use to conservation decisions that are needed *now*.

Framing Simple Questions

The question should be as free as possible of jargon and of technologies that require considerable expenditure and training. If it's not possible to phrase a conservation concern and the question that results in clear, understandable language, perhaps the question isn't so urgent after all. By using clear language and common names of plants and animals instead of Latin names, you may greatly increase your ability to involve local people in the full process of inquiry and conservation (see Cooperrider

1996; Margoluis and Salafsky 1998). For example, consider the pompous question "Do transient aggregations of semi-feral *Gallus gallus* associated with adjacent subsistence agricultural establishments negatively impact propagule survival and juvenile recruitment of native arborescent vegetation within the management module, as compared with control exclosure plots?" While this wording might impress writers of government documents, less pompous phrasings do exist. Can you suggest one? How about "Do chicken flocks that wander into the reserve eat many seeds and seedlings of native plant species and cause plant regeneration to differ from that in comparable chicken-free areas?"

Likewise, if answering the question requires expensive and training-intensive technology, in many cases—although certainly not all—it's possible to propose a similar alternative that depends primarily on the proper use of the most versatile tools of all: your eyes, brain (including its accumulated knowledge of natural history), and hands. I don't mean to disparage the appropriate use of technology during the inquiry process. Nevertheless, evaluating many of the most urgent concerns confronting conservation professionals depends more on proper application of the inquiry process than on the use of technological marvels (box 3.1). For example, consider the management concern

---

Box 3.1. GIS: Confusing Tools with Questions?

Conservation professionals as well as academic ecologists often tend to confuse their questions with their tools. A conspicuous example is an acronymic technique that's become the rage in conservation and management. GIS, or geographic information systems, are "computer hardware and software packages designed to store, analyze and display spatially referenced data; they deal with information that can be related to some form of map" (Haslett 1990). For example, numerous points in a landscape can be characterized in terms of elevation, slope, exposure, soil, land-use history, current land use, infrared signature, vegetation, and other physical, biological, or sociological variables. The data are entered into a computer program along with the precise geographic coordinates, followed by various processes of uncovering the relationships between the different variables, displaying overlays, and so forth (Wessman 1992; Miller 1994).

Despite its space-age aura, though, GIS technology is simply a fancy tool that, just like a carpenter's hammer, is useful for some tasks but not others. In itself GIS constitutes neither science nor management, just as a hammer by itself does not constitute building construction. The current rush by many managers of protected areas, aided and abetted by well-meaning staff in national ministries or international donor agencies, to invest heavily in GIS technology and the personnel to use it may be an unfortunate waste of time and resources unless there is a clearly defined framework of scientific inquiry and management concerns in which GIS technology would be a useful tool. Under many circumstances, of course, GIS approaches can be of tremendous value. For example, Lewis (1995) describes a highly appropriate use of GIS in tropical Africa that encompasses the points to be made in several of the later chapters of this book. Kiester et al. (1996) describe a sophisticated application to a North American setting, while Powell, Barborak, and Rodrígues (2000) use GIS for a country-wide analysis of the conservation status of all Costa Rica's life zones. The purchase of GIS instrumentation or even its prolonged use, though, won't help you to decide among conservation alternatives if you haven't decided on a question beforehand.

over the best way to enhance the recovery of a watershed affected by illegal gold mining (chapter 2). Some excellent questions generated by that concern—and complying with all the preceding criteria—would involve sophisticated technologies for analyzing soil chemistry and stream chemistry, almost certainly resulting in precise and objective answers. But with limited resources you might address the same concern by framing equally excellent questions that require straightforward and inexpensive methodologies such as sampling stream insects (chapter 8) or measuring the growth and survival of plant seedlings. Technology and technical language may increase the precision of your question and results, but add them only as necessary.

The question should be simple in another sense also. Limit the number of concerns or factors you include in a single inquiry. Especially if you're new to field work, you'll be tempted to explore all possible factors at once. For example, the thesis question of a very bright and eager student I know was something like "What's the effect of different combinations of substrate, type of container, amount of sun, type of fertilizer, watering schedule, place of origin, and planting season on the rate of germination of seeds of this species of medicinal herb?" Being only one person with a finite number of seeds at her disposal, she found it a bit difficult to come up with a definitive answer.

Framing Questions: Some Practice and Some Examples

Because skill at framing questions is the key to scientific inquiry and management, I strongly suggest that you go through the exercise of box 3.2 before reading any further.

Once more, in general terms, *all questions that are stimulated by observing one's surroundings are valid*. Still, some lead much more readily than others to firsthand inquiry and practical applications. Examples of "less than adequate" questions, by this measure, abound, and there's no need to list them here. Fortunately, examples of well-phrased questions also abound. I've culled a small sample of questions from the recent literature in applied ecology, all falling somewhere in the range between small-scale basic studies (figure 2.5) and urgent, large-scale conservation questions (figure 2.4). Sometimes the original idea wasn't explicitly presented as a question, so I've paraphrased the

---

Box 3.2. Practice with Framing Questions That Lead to Inquiries

If you went through the exercise of box 2.1, review the list of questions you generated. Carefully rephrase each one as necessary so that it's *answerable, comparative, alluring*, and *simple*. Make sure that you retain the basic concept or concern that stimulated the question originally as well as the spatial scale of that exercise.

Now practice framing "real-life" questions. Wander about outdoors, preferably in the protected area or other landscape where you work, and develop questions based on your real conservation concerns or on ideas for a possible field study. This time, of course, the questions could involve any scale of space and time, the only requirement being that each question should conform to the four criteria in this chapter. At the end of the exercise, review your questions—better yet, ask a colleague to review them—to make sure that there isn't an even better way to word them with respect to those criteria.

Figure 3.1.
A reserve border that experienced
selective logging in the past
(Napo Province, Ecuador).

author's writing a bit. In all cases, though, the idea conformed to most or all of the four guidelines discussed in the preceding sections. In all cases, results of inquiries initiated by these questions have already influenced, or could influence, the development of conservation and management guidelines:

1. Compared with unlogged forest, what is the effect of low-level, selective logging (figure 3.1) on bird assemblages (Thiollay 1992)?

2. Is there a substantial change in the local and regional diversity of native animals (e.g., birds, insects, spiders, lizards, earthworms, small mammals) when traditional shade-coffee plantations are converted to modern sun-coffee plantations (Perfecto et al. 1996)?

3. In the restoration of native vegetation on degraded lands, does setting out perches for birds change the arrival rate of viable seeds and thereby affect the recovery process (McClanahan and Wolfe 1993)?

4. What is the impact of trout introduction on the native fauna of cool-water streams, as compared to streams with no trout present (Flecker and Townsend 1994)?

5. Do populations of native tree species benefit more from the presence or absence of domestic livestock (Reid and Ellis 1995)?

6. Do roads and major trails affect the rate at which exotic, weedy plants invade natural areas (Tyser and Worley 1992)?[1]

7. In lowland wet forest reserves with dirt trails often used by visitors, is the runoff of surface water, and accompanying soil erosion, greater or lesser from trails than from the untrampled forest floor nearby (Wallin and Harden 1996)?

8. What is the relative physical impact that (a) horses, (b) llamas, and (c) humans have on trails and on rates of soil erosion (DeLuca et al. 1998)?

9. What is the relationship between numbers of hikers that pass along trails and the abundance or diversity of native birds (Riffell, Gutzwiller, and Anderson 1996; Miller, Knight, and Miller 1998)?

10. What is the present-day relationship between hunting intensity and population density of large vertebrates (Hill et al. 1997)?

11. Likewise, how do population densities of mammals vary among protected areas with different restrictions on hunting (Carrillo, Wong, and Cuarón 2000)?

CHAPTER 4

Design: Matching Data Collection to the Scope of the Question

If you worked in areas inhabited by demons you would be in trouble regardless of the perfection of your design . . . [still,] replication and interspersion of treatments provide the best insurance.

—Stuart H. Hurlbert (1984)

Now for the inquiry proper. Let's say that you're a conservation professional who has embarked upon the management cycle (figure 2.4) by articulating a critical conservation concern. For example, selective logging has been taking place within the boundaries of the multiple-use reserve where you work. It's imperative that you propose management guidelines (see figure 1.1). These might be to allow logging to continue at previous levels or to institute a total moratorium on logging. Or you might need to propose new limits on the maximum intensity of logging. In particular, you're concerned about the possible effects of logging on forest birds, understory frogs, and terrestrial small mammals because these are the animal groups for whose protection the area was set aside in the first place. So you frame the question, complying with the four criteria presented in chapter 3. At this stage the question might resemble the first one of the list given at the end of chapter 3. For example, let's begin with "Does selective logging significantly alter the abundance and diversity of forest birds, frogs, and small mammals within the reserve?"[1] Now, how will you answer the question so that the results can help you to decide on the management guidelines?

Your conservation concern and the question that follows have set boundaries on the scope of the study: a limited space (within the borders of the reserve) and probably a limited time as well (for example, one year at most before you must propose the guidelines). Even within that well-defined scope, though, you can't be everywhere all the time examining all possible responses of every bird, frog, and small mammal to past or present logging activities as compared to every individual bird, frog, and small mammal that lives in unlogged areas. All you can do is to *sample* a limited subset of those instances. You must *design* the layout of the study (see figures 2.4, 2.5) in time and space, selecting the sample that represents the best compromise between the most honest, complete answer (which would require you to be omniscient) and the time and effort you have available.

Using the question on selective logging as an example,[2] we'll now go through a scheme for design that emphasizes logic and a fundamental understanding of natural history rather than the nuances of statistical theory.[3] You'll see that successful design depends in large part on an appreciation for one of the most striking features of landscapes: natural variability, patchiness, or heterogeneity (and see chapter 6). In order to clarify key concepts here and in chapter 5, I'll need to introduce a fairly lengthy vocabulary (box 4.1). The vocabulary is intended to clarify rather than obscure

---

Box 4.1. An Informal Glossary for Chapter 4

Block design: A study design in which you locate a full set of response units at each of several different points, in space or time, with respect to potential confounding factors. Each set, or block, consists of one response unit per level of the design factor or (better) a consistent number of ≥ 2 such units per level.

Confounding factor: A phenomenon that's not included in your question but influences the values you record in a nonrandom manner. It remains just a *potential* confounding factor unless you fail to take it into account in your study design. In the latter case, unbeknownst to you the influence of the confounding factor on the values you record may be aligned with or against the influence of the design factor you're examining. In consequence, you might mistakenly attribute the patterns—or lack thereof—in your results to the design factor alone, not only in the statistical analysis but also in your reflection and application.

Design factor: As specified by your original concern and your question, the factor whose possible effect on the response variables you intend to examine and around which you design the inquiry. In some but not all cases the design factor can also be called a *treatment factor,* though that term is best limited to design factors with categorical levels (see "Levels of the design factor" below).

Evaluation unit: The standardized unit in which, or for which, you record the value of the response variable. In a few cases the evaluation unit encompasses much or all of the response unit (see below). In many cases the evaluation unit is much smaller and more narrowly defined than the response unit.

Levels of the design factor: The different classes or values that the design factor can assume and that you're examining in this inquiry. If only a few distinct classes exist, they are *categorical levels* and could also be called *treatments.* If the design factor can take on any one of a large number of values such that each one of the response units could have a unique level, they can be termed *continuous levels*—even if the number of possible levels is finite in practice—and often the design factor is later treated as an "independent variable" in statistical analyses.

Manipulative (experimental) study: A study in which you yourself assign the different levels of the design factor to the individual response units—which, in this case, can also be called "experimental units" (see "Response unit" below).

Nonmanipulative (nonexperimental, observational) study: A study in which you work with response units that already display different levels of the design factor or that experienced the different levels prior to your arrival on the scene.

Pseudoreplication: The sin of mistaking the different evaluation units within a single response unit for true, replicate response units themselves, as defined by the question and the design factor. For absolution, you must either change the design or change the question.

Question: What you propose based on the original concern or idea for a study, the use of a question mark, and the process of complying with the four criteria listed in chapter 3. The question must specify the scope of the study to follow. It must also specify, or refer clearly to, the design factor(s) and the general nature, though not the details, of the response variable(s) you'll measure.

Randomized design, randomized treatments design: A design in which you assign (in the case of a field experiment) or sample (in the case of an observational study) the response units for the different levels of the design factor in a random fashion with respect to any potential confounding factors. Presumably, this prevents the latter from metamorphosing into true confounding factors.

Response unit: The fundamental unit of study design and analysis, or the minimum individual unit that manifests the effects of the design factor(s) yet is independent of other such units with respect to the question. For example, if your question concerns differences in the rate of leaf litter decomposition between black-water rivers and white-water rivers (note the plurals), each river is a single response unit. If your question concerns differences in the rate of leaf litter decomposition between one particular black-water river and one particular white-water river, each independent site you sample within a given river is the response unit. In manipulative studies, the response unit can be termed the *experimental unit.*

Response variable: The property whose variation with respect to the design factor is the focus of your study and whose value you measure and record in every evaluation unit. If you're subsampling, to minimize the risk that you'll inadvertently commit pseudoreplication, you should characterize the entire response unit with a single value (for example, the mean or median among the values from the different subsamples) before proceeding with data analysis. Often the response variable becomes the "dependent variable" in the statistical analysis that follows.

Sampled population: The set of entities—for example, small mammals or plant seedlings—that your methodology is really sampling given that very few methodologies sample the target population (see below) without bias. The sampled population may closely approximate the target population that you wish you were sampling, or the two may be quite different

Box 4.1. Continued

without you realizing it. Note that this use of the term "population" is usually distinct from its use to describe a group of individuals of the same species.

Sampling methodology: The standardized technique by means of which you measure the response variable in each evaluation unit.

Sources of variation: The three possible causes of the variation in the value of the response variable among evaluation units or response units. The three sources are:

• *Intrinsic variation:* The variation among different observations that always exists in the absence of any other influence. After all, you'll never encounter two absolutely identical response units. Also called "error" or "sampling error," but intrinsic variation isn't error: it's reality.

• *Variation associated with the design factor:* The variation that's contributed by the effects of the different levels of the design factor.

• *Variation associated with confounding factors:* The variation that's contributed by the effects of the confounding factor(s). In poorly designed, analyzed, or interpreted inquiries, this is often confused with the variation associated with the design factor.

Subsamples: Multiple evaluation units within a single response unit. Each subsample provides one estimate for the value of the response variable within that response unit as a whole, but the set of subsamples as a whole provides an estimate that's much more likely to be representative.

Subsampling: Setting up a design and methodology that makes use of subsamples. Often this is useful or even necessary. At other times it's a waste of time and energy that could better be devoted to increasing the number of response units.

Target population: The set of entities—for example, small mammals or tree seedlings—that you intend to sample without bias. Your question defines the nature of the target population, and you should seek the sampling methodology most likely to minimize the discrepancy between the target population and the sampled population (see above).

the concepts involved, and to present a unified approach to the design of inquiries in the field. You won't find all these terms in traditional texts on design and statistics; some analogous terms you *will* find there are often inconsistent or used in inconsistent fashion. If you're bemused or put off by the complexity of this chapter, first have a go at appendix D, then return here and see if it doesn't make more sense. In any case, please note that every point in this chapter applies just as much to basic field studies (figure 2.5) as to questions on conservation (figure 2.4).

Some Alternative Designs Regarding the Logging Question

What are some designs for studies meant to answer the question "Does selective logging significantly alter the abundance and diversity of forest birds, frogs, and small mammals within the reserve?"

Whatever the design, you'll examine these three groups of vertebrates only in certain places and at certain times, under two different conditions: unlogged (UL) and selectively logged (SL). You have many choices of places and times for locating your standardized *evaluation units.* You also have several choices for *response variables,* in this case the alternative ways to record abundance and diversity of each animal group within each evaluation unit.

Here you'll need a different sort of evaluation unit for each animal group. For birds, let's say that the evaluation unit consists of a 1-km zigzag transect along which you walk slowly for one standard two-hour period (0600 to 0800). The two response variables you'll use to express abundance and diversity in each unit are, respectively, the number of individuals and the number of species of known forest-dependent birds seen or heard within a 50-m radius. Likewise, for frogs of forest understory the two response variables will be numbers of individuals and numbers of species encountered in the stratum from 0 to 2 m above the ground, along a belt transect 250 m in length and 2 m in width that's surveyed carefully between 2100 and 2300. Response variables for small mammals will be total numbers of individuals and species of known forest-loving mammals captured per evaluation unit, a 20×20 grid (with 10-m spacing) of 400 Sherman live traps opened on the ground for one night at the site.

The reserve includes three major habitats: a region of lowland tropical wet forest, a mountain range covered by cloud forest, and a region of tropical dry forest in the rain shadow of the mountain range (figure 4.1). Each habitat includes several tracts where forestry concessions were granted, and

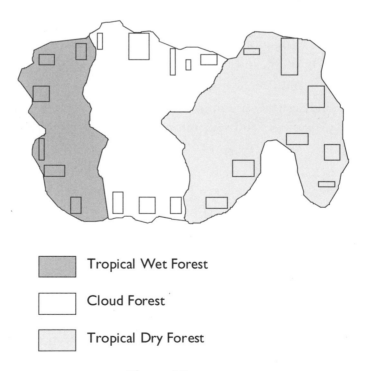

Tropical Wet Forest

Cloud Forest

Tropical Dry Forest

Figure 4.1.
A hypothetical large reserve in which selective logging has recently taken place. Rectangles within the reserve indicate selectively logged (SL) tracts, with sizes greatly exaggerated for purposes of illustration. The remainder of the reserve, the vast majority of its area, is currently unlogged (UL) but could experience selective logging in the future, either as part of the study design (e.g., design 13) or following the development of management guidelines.

selective logging took place, in the recent past. Future logging, if any, will take place in concessions within those expanses of forest that are still UL at present. The size of any evaluation unit is small relative to the size of an SL tract or a continuous expanse of UL forest, so that you could feasibly locate from one to several different evaluation units within a given tract. Scrutinize figure 4.1 and the following alternative designs, some of which figure 4.2 illustrates. *Before reading the next section,* choose one or two designs from the list below that you feel best fit the question as asked. In designs 1–6 you compare two evaluation units per animal group, in designs 7–13 you compare twelve evaluation units, and in designs 14–16 you make four distinct visits to each of twelve evaluation units. So, the question is, where will you place the evaluation units (and when will you visit them)?

DESIGN 1. You locate one evaluation unit per animal group at one site in one tract of *SL tropical dry forest* at 300-m elevation above sea level on the E slope of the mountain range, and one evaluation unit per animal group at one site in *UL cloud forest* at 2,000-m elevation on the W slope (figure 4.2a).

DESIGN 2. You locate one evaluation unit at one site in one tract of *SL tropical wet forest* at 300-m elevation on the W slope, and one unit in *UL cloud forest* at 2,000-m elevation on the W slope.

DESIGN 3. You locate one evaluation unit at one site in one tract of SL cloud forest *at 1,600-m elevation,* and one unit in UL cloud forest *at 2,000-m elevation* on the same slope.

DESIGN 4. You locate one evaluation unit at one site in one tract of SL cloud forest *on flat terrain,* and one unit in adjacent UL cloud forest *on a 30° slope,* both sites at approximately 2,000-m elevation.

DESIGN 5. You locate one evaluation unit at one site in one tract of SL cloud forest at 2,000-m elevation on a 30° N-facing slope, and one unit in UL cloud forest at 2,000-m elevation on the adjacent 30° S-facing slope.

DESIGN 6. You locate one evaluation unit at one site in one tract of SL cloud forest at 2,000-m elevation on a 30° N-facing slope, and one unit in adjacent UL cloud forest *on the same slope* (figure 4.2b).

DESIGN 7. As in design 1, but with twelve evaluation units. You locate one at each of six sites scattered randomly *within the one SL tract in lowland dry forest,* and one at each of six sites scattered randomly *within a similar-sized region of the UL cloud forest* (figure 4.2c).

DESIGN 8. As in design 1, but with twelve evaluation units, one at each of six sites. You locate each in a different SL tract chosen at random from among all SL tracts *in the lowland dry forest,* and one at each of six sites chosen at random from all possible areas *of the UL cloud forest* at 2,000-m elevation (figure 4.2d).

DESIGN 9. As in design 6, but with twelve evaluation units. You locate one at each of six sites scattered randomly *within the one tract of SL forest* and one at each of six sites scattered randomly *within a similar-sized region of the adjacent UL forest.*

DESIGN 10. As in design 6, but with twelve evaluation units: instead of one pair of evaluation units in adjacent SL and UL tracts, you locate six such pairs, *each pair involving one site in an SL tract and one site from the UL forest alongside,* the pairs being scattered across the full variety of slopes and exposures available *in cloud forest at 2,000-m elevation.*

DESIGN 11. You locate twelve evaluation units, one from each of six sites selected randomly *from among the full range of SL tracts throughout the forest types,* slopes, elevations, and exposures present in the reserve, and one from each of six sites also *selected randomly from throughout the UL areas* (figure 4.2e).

DESIGN 12. You locate twelve evaluation units in a combination of designs 10 and 11: six pairs of units, each pair involving one site in an SL forest tract and one in the UL forest alongside, the two sites matched as closely as possible with respect to all other characteristics and *the pairs being scattered throughout the full range of forest types, slopes, elevations, and exposures present* in the reserve (figure 4.2f).

DESIGN 13. As in design 12, but *an experimental (manipulative) study* as opposed to an "after the fact" nonexperimental (observational or nonmanipulative) study as in all preceding designs. You yourself select six areas at random, each area being *twice as large as* the typical SL tract, from throughout *the currently UL parts* of the reserve. By flipping a coin you randomly assign one half of each to be newly opened to selective logging and the other half to be left unlogged, then complete the study at a certain time after logging has taken place.

DESIGN 14. As in design 12 or design 13, but you sample the six units in UL forest each four times, *all during the wet season,* and the six units in SL tracts four times *during the dry season.*

DESIGN 15. As in design 12 or design 13, with each of the twelve units, UL and SL, sampled four times *during the wet season only.*

DESIGN 16. As in design 12 or design 13, with each of the twelve units, UL and SL, sampled *twice during the wet season* and then again *twice during the dry season.*

Sources of Variation among Results

Let's assume that you follow design 1 (figure 4.2a) and record, for example, forty-six forest bird species along the two-hour transect that's the evaluation unit at the site in the UL forest and thirty-nine forest bird species in the evaluation unit at the site in the SL tract. Is this good evidence that selective logging has a significant, detrimental effect on the diversity of forest birds and that you should immediately propose a complete moratorium on logging throughout the reserve? After all, the result—fewer birds in the SL site than in the UL site—confirms your fears and conforms to conventional wisdom. To answer this, let's examine the possible sources of variation among the data collected in this or any other study.

Variation That Exists No Matter What

First of all, the difference between the two values might be due to chance alone. After all, how likely is it that you'll obtain identical values of a given response variable for any two evaluation units? In this case, how likely is it that you'd record precisely the same number of bird species along two different transects even if both were located in the same forest tract or exactly the same transect were walked twice on successive days? Not very likely.

So, it's highly likely that different evaluation units will produce different values even under iden-

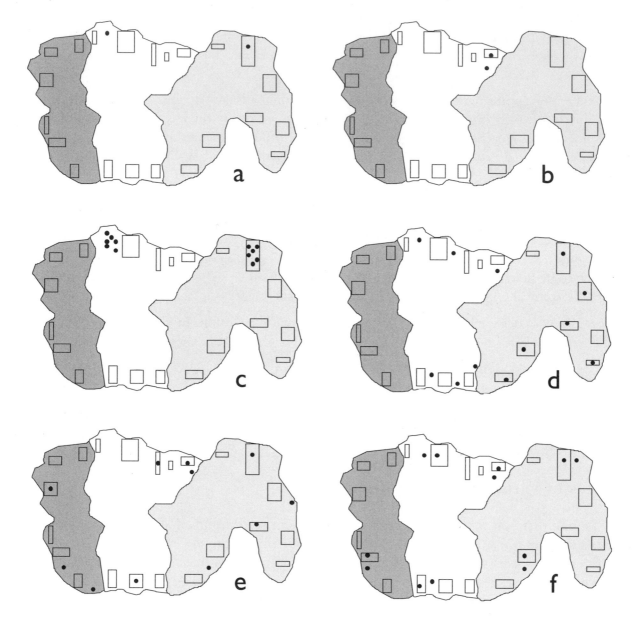

Figure 4.2.
Some alternative designs for studies to address the concern about selective logging. On each map, dots indicate sites of the two or twelve evaluation units. See figure 4.1 for interpretation of the map, and see the text for a full description of the designs. a: design 1; b: design 6; c: design 7; d: design 8; e: design 11; f: design 12.

tical conditions. You have two such units in hand. One happens to fall in an SL parcel, the other in a UL tract. If there were absolutely no differences between the underlying bird assemblages in those two regions but by chance alone the evaluation units themselves yielded different values, doesn't it follow that one value would have to be larger than the other? This time you recorded the larger value in the site that happened to fall in the UL tract. What's the chance of recording the larger of the two values at one site as opposed to the other? Exactly 50 percent, the same as the chance that a flipped coin will land heads up instead of tails. You wouldn't use one coin flip as evidence that the coin is biased toward landing one way or another. Likewise, just because the evaluation unit in which you recorded the greater number of bird species happened to exist in a UL site and the unit with the lower number in an SL site doesn't mean that you can confidently blame selective logging for the difference in numbers of bird species, no matter the magnitude of that difference, *unless you can demonstrate that said magnitude exceeds the normal range of intrinsic variation in counts of bird species among evaluation units.* And you have no idea of the nature of intrinsic variation among units because you haven't examined *replicate* SL and UL sites to see how much they vary among themselves discounting any effect of logging.

Some following sections and chapter 5 discuss the concept of intrinsic variation in much more detail. For now, just recognize that intrinsic variation exists among evaluation units and among response units (see step 6 of "Designing an Inquiry . . ." below), even in the absence of any other influence that affects what you're measuring in each. Thus, you cannot automatically ascribe an observed difference in these values to any other cause until you have taken the possibility of intrinsic variation into account. No matter how much effort you put into a single evaluation unit (say, for birds a zigzag transect of 80 km walked for ten consecutive days) or how reasonable sounding the comparison (say, that of design 6, figure 4.2b), each of designs 1–6 involves only two evaluation units. One will surely display a higher value than the other, for any given response variable. Is selective logging responsible, or is it intrinsic variation among evaluation units or even that among different forest tracts, regardless of logging history? Not one of designs 1–6 can resolve this serious doubt.

Variation Associated with the Factor You're Examining

Selective logging is your *design factor*. It's the subject of interest or concern, it's specified in your question, and it's the phenomenon around which you design the study. Naturally, it's always possible that the values you obtain do indeed reflect some influence of the design factor. Here, it's entirely possible that one underlying cause of the observed difference in bird diversity (forty-six vs. thirty-nine) between the two evaluation units is the influence of selective logging in the latter case. That's what you'd like to know. But with only these two values you have no way of knowing it.

Variation Associated with Factors You're Not Examining

It's also possible that a great deal of variation in the values you measure among different evaluation units or entire forest tracts results from features that you never intended to investigate, features that your question and design failed to take into account. That is, the design, the results, and your interpretation of them may be confounded by other, unacknowledged influences that also vary among the units and are aligned with, or opposite to, the influence of the design factor. That is, the influence of *confounding factors* on the values you're measuring might exaggerate, contradict, obscure, or otherwise complicate the effect that the design factor would have on your results were it acting alone.

For example, return to design 1 (figure 4.2a) and ignore for the moment the influence of intrinsic variation. The SL sampling site sits in tropical dry forest, the UL site in montane cloud forest. In addition to the intended design factor (selective logging), do the two sampling sites differ in any features that might have influenced bird diversity? Obviously. For example, tropical dry forest and high-elevation cloud forest clearly differ tremendously in mean rainfall and temperature, physical structure, availability of food resources, and the seasonality of all four features. Any or all of those factors could certainly exert profound effects on bird assemblages, contribute to the difference between the species counts of forty-six and thirty-nine, and confound any influence that selective logging alone might have on those counts.

Are other designs less confounded? Because tropical wet forest and cloud forest share some features in common, the confounding effects of rainfall patterns and forest structure in design 2 may be less intense than for design 1. Nevertheless, other confounding factors (differences between the sites in forest structure, temperature, food resources, and many other features) remain in force. Although the two sites in design 3 are both in cloud forest, the 400-m difference in elevation implies substantial differences in both physical and biological features. It follows that the two avifaunas will almost certainly differ in any event. In design 3 you still can't blame an observed difference in the number of bird species solely on the presence of selective logging in one site but not the other. In design 4, the difference in slopes between the two sites implies differences in forest structure, food availability, and avifaunas. The forest on the steep slope almost certainly has a more open canopy, greater light penetration, and thicker understory than the forest on flat terrain. In design 5, the different exposures of the two sites likely lead to differences in at least cloud cover, wind, and moisture—hence in the local avifaunas even in the absence of selective logging. So much for designs 1–5.

At first glance design 6 (figure 4.2b) seems free of confounding factors. In fact, many published field studies have had similar designs (in addition to numerous published studies whose setups resembled designs 5, 4, 3, 2, and even 1). Nevertheless, many subtle factors besides logging history could differ between the two side-by-side forest tracts. For example, perhaps one tract experienced more landslides in the past century than the other, or there's a spring on one site but not the other, or the topographies and therefore the forest structures differ—or before logging, the SL site held a greater density of commercially valuable timber species than the UL site, a very likely possibility. The present-day avifauna might reflect those or other confounding factors at least as much as the effect of selective logging. In short, as you proceed from design 1 through design 6 and the potential role of confounding factors changes from the obvious to the subtle, you may be tempted to feel more and more comfortable with the design's validity. You shouldn't.

More complex designs, those that appear to take intrinsic variation into account, don't necessarily help with confounding factors. Look at design 7 (figure 4.2c) versus design 1, or design 9 versus design 6. Designs 7 and 9 involve several evaluation units in each of the two forest tracts concerned, but all that accomplishes is to give you a better idea of a representative value for each tract as a whole. There's still only one unique forest tract to represent each of the two conditions, UL and SL. You still have no way of evaluating the amount of intrinsic variation in each measure among different SL parcels and that among different UL tracts. The chance that the value you record for each of the two forest tracts as a whole reflects the influence of confounding factors, not just the design factor of selective logging, is just as great in designs 7 and 9 as in designs 1 and 6, respectively.

Design 8 (figure 4.2d) does allow you to evaluate the extent of intrinsic variation in bird counts among different, independent tracts. Design 8, though, suffers seriously from the confounding effects of the two very different forest types in which all the SL and all the UL sites respectively occur (see figure 4.1). Design 14 includes a rather different confounding factor—season—that's aligned with the design factor and could greatly influence the observed abundance and diversity of all three animal groups, especially birds and frogs. Under "Designing an Inquiry . . . " (see below), we'll return to the designs and to the questions they can or can't legitimately answer. Meanwhile, though, among designs 1–12 it's beginning to look as if only designs 10, 11 (figure 4.2e), and 12 (figure 4.2f) might be adequate.

What to Do

While designing your study, be as sure as you possibly can that you're really investigating the influence of *design factors* on the features of interest and that you can disentangle them to the extent possible from the influences of *intrinsic variation* (see chapter 5) and *confounding factors* (see step 9 in the next section). If it's too late—that is, if the inquiry (your own or someone else's) is already under way or completed—at least reflect long and hard on whether it has truly addressed the influence of the design factor; then draw your conclusions or plan your applications accordingly.

Designing an Inquiry in Eighteen Steps

The chapter title says it all. Either adapt the design to the scope of your question or adapt the question to the scope of the design. Indeed, at every step of the design process, you should reexamine the question to see if it's compatible with the design and, if not, rephrase it as necessary. The reason: as you proceed you'll discover additional logical or biological subtleties that must be taken into account by modifying either the design or the question. In fact, you should . . .

Step 1. Scrutinize the Question at the Outset

Until now the question has been "Does selective logging significantly alter the abundance and diversity of forest birds, frogs, and small mammals within the reserve?" Is that precise enough? What's the scope of your management concern? If logging has occurred and will continue to occur (the eventual guidelines permitting) only along the park's perimeter, are you truly concerned about the effects of logging throughout the entire reserve including its core? No. So, you'd better reword the question right away. How about "Does selective logging *in the periphery of the reserve* significantly alter the abundance and diversity of birds, frogs, and small mammals there?"

Step 2. Decide Between Doing an Experiment and Sampling from What's Already There

Designs 1–12 are observational or nonmanipulative studies. You compare evaluation units from SL and UL sites, but you've arrived after the fact: you yourself weren't responsible for deciding which tracts would be logged and which areas would be left unlogged. Design 13 is a manipulative or experimental study:[4] you decide on the treatment (SL or UL) that each forest tract—and thus each sampling site within each tract—will receive. What's the difference?

In Design 13 the only criterion used to determine whether the left- or right-hand tract will be SL is the flip of a coin. Presumably, such randomization will minimize the chance that one or more confounding factors is consistently aligned with the design factor. You stand a reasonable chance of truly observing the influence of selective logging on the results. Design 12 (figure 4.3f) is not an experimental study. Strong though it seems, design 12 runs the risk of including at least one cryptic confounding factor that's still aligned with the SL-UL contrast: the behavior of loggers. What if past loggers selected their tracts on the basis of certain features that, independently of the logging itself, also had influenced abundance and diversity of preexisting birds, frogs, and small mammals— for example, the density of marketable trees or the ease of their extraction? Even before logging, then, the fauna of the soon-to-be-SL tracts as a group might have differed from those areas bypassed by the loggers, the expanses of UL forest. You'd be wrong if you attributed all post-logging differences between the faunas of UL and SL tracts to the effects of the logging alone.

In short, among designs 1–13, only design 13, or a similar experimental design, can truly answer the question as it was originally phrased, on the *effect* of selective logging. Designs 10–12, or even stronger nonmanipulative designs, can't quite answer that question. Instead, designs 10–12 answer a different one. In those designs you're really investigating present-day differences between tracts that were either logged or not in the past. The question becomes "Along the perimeter of the reserve, *is there currently a significant difference* in faunas between SL sites and UL sites?" If you can't perform a manipulative study (design 13), just rephrase the question and perform the strongest possible nonmanipulative study. During the reflection and application steps (see figure 2.2), keep the appropriate wording for the question in mind so that a persistent doubt will nag you every time you're tempted to state unequivocally that logging alone was directly responsible for the present-day differences you observe between sites that were logged and sites that were not logged, before you arrived on the scene.

Does this mean that you should always prefer an experimental design to an observational one? By no means. Each class of designs has its advantages and disadvantages. The real advantage of observational studies is that they involve the situations and conditions that occur "naturally." The chief disadvantage of observational studies is their inability to control for all possible confounding factors. They can tell you about the way results vary among units that present different aspects of the design factor, but they can't tell you whether those results came about through the agency of that design factor alone. Conversely, the advantage of experiments is that they begin in the present, not the past, and you're in control from the start. Thus, well-designed experiments allow you to answer, with reasonable confidence, questions about effects of A vs. effects of B, not just about present-day differences among sites that experienced A vs. B in the past. Some disadvantages, though, are that experiments may be logistically difficult, impractical, costly, or clearly impossible, especially when they involve a large scale in space or time. Another serious concern—ethics—will be discussed in detail below. Finally, the cleverness of a well-designed and well-controlled experiment may be misleading: because it does not involve situations and conditions that occurred "naturally," an experiment may actually be less realistic, with respect to the conservation concern and the development of guidelines, than a well-designed observational study (see also Camus and Lima 1995).

For example, in this study you're most concerned about the possible effects of real-life logging carried out by real-life loggers in the future. If you follow experimental procedures and randomly select the tracts yourself, you're not only controlling for but also ignoring the criteria by which loggers select tracts. Given that loggers, not you, will be the ones to select tracts in the future, through good experimental design you're unknowingly investigating an artificial situation and perhaps drawing erroneous conclusions with respect to reserve management. One reasonable option is to continue with the experimental design 13 but involve loggers at all steps in the design, particularly in the selection of the "double-wide" experimental areas. You'd still control, by flipping your coin, which half of each area would be logged. Nevertheless, if you're confident that future loggers will use criteria similar to those used by past loggers in their choice of tracts, the other option is to revert to a nonexperimental design, say design 12 (figure 4.2f), and rephrase the question as proposed above: "Is there currently a significant difference . . . ?"

Step 3. Specify the Scope of the Question in Space and Time, Making Sure You Can Sample Throughout That Scope

In the first two steps, you've already adjusted and readjusted the original question. If you're concerned about logging throughout the entire perimeter of the park, then you must sample from throughout that perimeter. In this sense any one of designs 11–13 is adequate. Design 10 or a corresponding experimental design is not. While an inquiry that follows design 10 may reasonably evaluate the question of differences between UL and SL sites, it does so only for the cloud forest habitat (see fig. 4.1) and really only for that portion of cloud forest that lies at about 2,000-m elevation. The scope of the study is much narrower than that of the question. What if you followed design 10 and then applied what you'd learned in cloud forest to proposing logging guidelines for the reserve's other forest types as well? It could easily be that whatever pattern you found in cloud forest is actually nonexistent or even reversed in lowland dry or wet forest—but you wouldn't know that. What if design 10 is your only practical alternative, though? Rephrase the question! For example, "*In cloud forest at 2,000-m elevation* along the perimeter of the reserve, is there currently a significant difference in faunas between evaluation units at SL sites and those at UL sites?"

Likewise, design 15 controls nicely for the potential confounding factor of seasonality—by eliminating it from the scope of the design. What question does design 15 really answer? If you're doing an observational study otherwise similar to design 12, it's: "*During the wet season of year X,* along the perimeter of the reserve is there currently a significant difference in faunas between evaluation units at SL sites and those at UL sites?" If you're doing an experimental study otherwise resembling design 13, it's: "*During the wet season of Year X,* along the perimeter of the reserve is there an effect of selective logging on the faunas?" Is either question really adequate to address your conservation concern? After all, many birds, especially fruit or nectar consumers, move in and out of habitats seasonally, so that comparisons between SL and UL sites could vary greatly with season. Likewise, frogs are notoriously seasonal and asynchronous in their appearances along census routes. Activity patterns and local abundance of some small mammal species may display marked seasonality. Therefore, to answer the original question in a way that is meaningful to management decisions, the study should be designed to sample from the scope of at least one annual cycle. In that sense design 16 is vastly superior to design 15.

Step 4. Specify the Design Factor

You're concerned about the possible effects of selective logging. That is, the question identifies selective logging as the design factor, whether the study is observational or experimental. Make sure that instead you haven't inadvertently designed the study around the factor "place"—unless that's what you want—or, even worse, the factor "evaluation unit." In reality, the design factor in designs

1–6 (figure 4.2) is "evaluation unit," for all these designs simply involve comparisons of two different evaluation units, one of which by chance happens to fall within a UL expanse of forest and the other within an SL parcel. Designs 7 and 9 include several evaluation units per forest tract but again only one tract each of SL and UL forest. They are adequate for answering the question "Is there a significant difference in faunas between these two forest tracts, one of which happens to be SL and the other UL?" That's all.

During the reflection phase (see figure 2.2) of a study in which you'd discovered a large difference between the faunas of the two forest tracts, each at a different location, you might cautiously speculate on the *possibility* that selective logging could have contributed to that difference. You really haven't *examined* that possibility, though, because designs 7 and 9 lack replicate examples each of UL and SL tracts, respectively. Design 8 does include various examples of UL and SL tracts, but all the SL sites are in one forest type and all UL tracts in the other. Therefore, what you're actually examining in design 8 is "forest type + selective logging," not selective logging alone. You'll run quite a risk if you use the results of any study following designs 1–9 (or 14, whose covert design factor is really "season") to propose conservation guidelines for selective logging practices.

Step 5. Specify and Justify the Levels of the Design Factor

If you've complied with the criteria of chapter 3 and your question is a comparative one, not only the design factor but also the *levels* you're examining should be clear. For example, here the levels of the design factor (selective logging) are "some" and "none." The better designs we've discussed so far include several replicate units (response units, as defined in the next section) for each of the two levels chosen. The levels are *categorical*. That is, either a site is UL, or it is SL. Can you justify having chosen these particular categories? Choice of the level "UL" is obvious—it's zero logging, the "control" condition. The level of "SL," for example in designs 10–12, can be justified if you've chosen a representative sample from the many extant SL tracts. In the experimental design 13, though, if you were to log less intensively on the experimental plots than is normal for SL tracts in the reserve, you might underestimate logging's true effects on birds, frogs, and small mammals. Conversely, if your enthusiasm (or that of the loggers you've contracted) to do a rigorous experiment got the best of you and you logged more intensively than is typical, you might overestimate logging's real-life effects and propose unreasonably strict management guidelines.

If you design a study with categorical levels of the design factor, make sure the levels represent typical conditions found in nature. Of course, the design doesn't have to be restricted to two levels. Here, for example, in addition to the "control" (UL), you might choose three levels of different logging intensities: light, medium, and heavy. Or, you might design a different study with levels as follows: UL; SL with mechanical skidders; SL with oxen; and SL with logs hauled out by hand, by teams of the loggers themselves (yes, this happens). Data from any study whose design factors have categorical levels are often evaluated statistically with a technique called analysis of variance (ANOVA for short).

Studies with categorical levels of the design factor, such as designs 10–13 or the four-level design just discussed, have many advantages: You can investigate more than one design factor at the same time,[5] the studies may be relatively easy to implement and interpret, and their power to detect effects of the design factor (see chapter 5) is often impressive. Their great disadvantage, though, is that you're choosing only certain levels. It could well be that in real life the design factor exerts its most notable effects at levels outside the range of, or in between, those you've chosen (Umbanhower 1994; Camus and Lima 1995). What's an alternative?

The alternative is to reword the question and choose a design in which the level of the design factor varies among response units such that each unit can display, at least in theory, a unique level. Levels of the design factor are now *continuous* for all intents and purposes, not categorical. Here one rewording of the question among many possible might be: "Along the perimeter of the reserve, what is the nature of the relationship between intensity of selective logging on the one hand and the abundance and diversity of forest birds, frogs, and small mammals on the other?" A nonmanipulative study to answer this question might involve twelve forest tracts or parcels selected at random from throughout the reserve's perimeter and displaying a range of past logging intensities from none (UL) to quite intensive. Upon arriving at a particular tract, you would first assess the logging intensity along some objective scale. This might be, for example, the number of trees logged per hectare at that site or the proportion of the tract's soil that was disturbed by logging activities. The value is the site's unique level of the design factor. Then, as before, you set up the evaluation unit(s) in that tract and record the numbers of species and individuals for each vertebrate group. Alternatively, you could design an experimental study with continuous levels—for example, by selecting at random twelve UL tracts and assigning to each a different intensity of logging to be carried out, including one tract with none at all. Whether the study is experimental or not, results are often analyzed with the class of statistical techniques known as regression analyses.[6]

Which should you choose, a design with continuous levels or one with categorical levels? The choice may depend on (1) the nature of the conservation concern, (2) the natural pattern of variation in the design factor, (3) the complexity of your question, and (4) the time and effort available for sampling.

1. *The nature of the conservation concern.* Your goal might not be to find a simple yes or no answer to the logging questions but rather to pinpoint the intensity of logging beyond which the abundance and diversity of any vertebrate group drops precipitously. You'll use this information in proposing guidelines on the maximum allowable intensity. In this case you'll most likely prefer a design with continuous levels.

2. *The natural pattern of variation in the design factor.* If selective logging has been and will be quite variable in its intensity, then again an approach using continuous levels may be most useful. In contrast, if selective logging occurs at one or a few quite consistent intensities, then a design employing categorical levels might be most appropriate.

3. *The complexity of your question.* If you intend to examine two or more design factors at the same time (say, logging intensity and number of years since logging occurred), or if the best way of dealing with confounding factors is through a block design (see step 9 below), then designs with categorical levels are more easily analyzed and interpreted than designs with continuous levels.

4. *The time and effort available for sampling.* However, complex categorical-level designs may require inordinately large numbers of response units, as defined in the following section. You might decide to stick with an easily implemented continuous-level study, controlling for confounding factors either by overwhelming them with brute force or by reducing the scope of the question (step 9 below).

For simplicity's sake, the remaining discussions in this chapter, and most in chapter 5, will continue to refer to designs with categorical levels such as those above. Nevertheless, please recognize that parallel considerations apply to the corresponding continuous-level designs.

Step 6. Specify the Response Unit

In a design with categorical levels, to distinguish variation associated with the design factor from intrinsic variation, you need several different, replicate cases, independent of one another, of each level. In a continuous-level design, you need a variety of such independent cases displaying a wide range of values for the design factor. If the design is manipulative, these independent cases are often termed *experimental units*. In a nonmanipulative study, they're sometimes called *samples*. The word *sample* has several other uses in study design and statistics, though. I prefer the term *response unit*, which signifies that each unit is "responding" independently from other such units to the influence (if any) of the design factor.

Now refer to steps 1, 3, and 4 above. *With respect to the selective logging question,* are multiple evaluation units that are located within a single SL parcel or a single UL tract, as in design 7 (figure 4.2c) or design 9, independent of one another? No. The different sites within a single tract all experience the same unique conditions of that tract today and/or experienced the same unique episode of selective logging in the past. The values you'll obtain from them, which will undoubtedly vary somewhat, are simply multiple measurements of the same responses of the same assemblage of vertebrates to the same unique set of conditions or the same unique episode. The six measures from the different sites within a tract may have their uses (see step 13 below), but assessing the effect of logging throughout the reserve's periphery—or assessing the differences between independent SL parcels as a group and independent UL tracts as a group—isn't one of them.

In contrast, the six UL tracts and six SL parcels examined in designs 11 and 12 are independent of one another, as are the six "double-wide" parcels of design 13. Here each tract or parcel is an independent response unit with respect to the question, whether the question involves effects (design 13) of selected logging carried out in the present, or differences among extant UL and SL tracts (designs 11, 12). The number of evaluation units you choose to place within each response unit (step 13) has no bearing on the definition, or number, of the response units. In other words, designs 7 (figure 4.2c) and 9, like designs 1–6, have only a single response unit per level of the design factor "selective logging." Whatever their other problems, designs 8 (figure 4.2d) and 10–13 (figures 4.2e, 4.2f) have six per level.

Step 7. Ensure That You Have True Replication among Response Units at Each Level of the Design Factor

This step, and this step only, enables you to distinguish variation associated with the design factor from intrinsic variation among different response units (Hurlbert 1984). Likewise, in order to distinguish variation associated with the design factor from that associated with obvious confounding factors, you must go to the next step.

Step 8. Intersperse the Replicate Response Units of the Different Levels—If This Makes Sense

Again, design 8 (figure 4.2d) provides six replicate response units per level of the design factor, but what's wrong? Refer to the discussion of confounding factors in the preceding section. In this design, all the UL response units are in one habitat and all the SL units in another. Just having a lot of replicate response units doesn't disentangle the design factor from the glaring confounding factors, due to differences in forest type, with which it's aligned (see Hurlbert 1984). Will design 11 (figure 4.2e), for example, do a better job of disentangling? Yes. The same reasoning should lead you to greatly prefer design 16 over design 14. After all, in design 14 the evaluation units, and the response units, aren't interspersed with respect to season.

In some studies it's impossible to intersperse the response units. Consider the following perfectly legitimate questions:

- In design 9, which has the greatest diversity of frogs per evaluation unit, the SL parcel or the UL tract nearby? (Note that with the change in the question, the evaluation unit becomes the response unit as well.)

- Which have the greatest diversity of forest birds, UL expanses of tropical wet forest or UL expanses of cloud forest?

- In SL tracts in tropical dry forest, at which season, wet or dry, is small-mammal density the highest?

- Does the onset of reproductive maturity of grizzly bears differ between populations in northwestern Canada and populations in Yellowstone National Park?

The design factors are, respectively, places, geographically separated habitats, seasons, and populations. All the response units for a given level occur grouped together in space or time, and I dare you to try to intersperse them—especially the grizzly bears. For simplicity's sake I'll continue to discuss questions, and designs, where interspersion is indeed possible. Please be aware, though, that you can ask a valid and important question regarding places or times and that, in such a case, your frustrated attempts to comply with the interspersion rule may drive you to madness.

Step 9. List All Potential Confounding Factors and Decide How to Deal with Them

Once you've listed all the potential confounding factors, your options are to:

1. randomly intersperse your response units for the different levels of the design factor with respect to potential confounding factors, using brute force to overwhelm the insidious effects of the latter on the data;

2. control for the most obvious of the potential confounding factors by restricting your sampling;

3. *block* across the most obvious potential confounding factors;

4. perform a manipulative (experimental) study; or,

5. live with the confounding factors and reword the question so that they no longer confound.

Here, in design 11 (figure 4.2e) you've taken the first option. You've randomly distributed SL and UL response units across the landscape of concern. This process has interspersed the response units with respect to many potential confounding factors, especially those associated with the great differences among forest types. The drawback, of course, is that now the results from the different UL response units will vary tremendously among themselves, as will those from among the different SL tracts. That is, to the intrinsic variation you've just added the variation associated with the potential confounding factors, in this case the different forest types. You can't distinguish the two, so you have to treat the "noise" as if it were all intrinsic variation. You'll be hard put to discern any influence of the design factor alone, if that influence is at all subtle. By increasing the number of replicate response units you may eventually mitigate this problem. The number necessary may be astronomical, though, if the design factor acts subtly on the values you're measuring while the potential confounding factors act flamboyantly (see chapter 5 and appendix B).

In design 10, you've taken the second and third options together. By restricting the study to cloud forest (option 2), you're controlling for those flamboyant, potential confounding factors associated with the great contrasts between different forest types (figures 4.1, 4.2d). Also, within the cloud forest SL and UL sites are *blocked* (option 3) with respect to location. That is, you've paired an SL tract with a UL tract at each of six different locations, themselves scattered across the cloud forest. This pairing makes it probable that the SL tract and UL tract of a given block experience very nearly the same effects of any potential confounding factor that varies from place to place within the cloud forest itself. The great benefit of blocking (option 4) is that your statistical analysis can now separate out neatly the variation among response units that's due to the design factor, from that due to those potential confounding factors that vary among the different blocks.[7] Effects of the design factor, if any, will be much easier to discern than in an unblocked design. Regarding option 2, the benefit of narrowing the scope of design 10 to cloud forest only is that you now have a more precise study than in design 11 and won't need as many replicate response units in order to discern patterns associated with the design factor. The great disadvantage of restricting the design's scope, of course, is that you've thereby restricted the scope of the question, the conclusions, and the applications you can reasonably make.

With respect to the original perimeter-wide scope of your concern and initial question, then, design 12 (figure 4.2f) obviously outdoes design 10, even though design 12 may require greater effort (travel time) than the latter. Design 12 also outdoes design 11 (figure 4.2e) because not only does design 12 extend over the scope of the question as does design 11 but also SL and UL response units are blocked, as in design 10. Again, blocking will enable your statistical analysis to distinguish the quantitative influence of the design factor on the response variable from the quantitative influence of being in a particular point in the landscape. With respect to season, similar reasoning applies to designs 15 and 16. Design 15 controls for the confounding factor "season," simply by restricting the scope to a single season. This restriction may enable design 15 to provide you with a very precise answer—but to a very restricted and rather useless question. In contrast, in essence design 16 blocks across time as well as space.

We already discussed the advantages and disadvantages of the fourth option, a field experiment. Here, in theory, design 13 is exceptionally strong, for it is not only experimental but also involves a block design, the UL and SL response units being side by side. The random assignment of the SL treatment to one of the two response units enables you to control not only for obvious confounding factors but also for subtle ones. You must decide for yourself, though, whether the degree of control imposed by an experiment is a good thing or a bad thing, with respect to the real-life phenomena your conservation guidelines or study ideas are meant to address.

Finally, for various reasons you may be unable to take any of the first four options presented above. Perhaps, for some reason or another, design 8 or design 14 is the only option available to you. If such is the case, simply reword the question and reflect carefully after the study. For design 8 the question might be some version of "Is there a significant difference in fauna between SL sites in lowland dry forest and UL sites in cloud forest?" Based on your knowledge of natural history in general and these two forest types in particular, you'll decide afterward just how far the results might apply toward resolving the original management concern. What might be an appropriate rephrasing for design 14?

Step 10. Select the Response Variable(s)

Now that you've decided on the nature of the response units and their layout, what is it that you'll measure in each one, or more precisely in each evaluation unit within each response unit? And are you measuring what you really intended to measure? These two all-important questions receive little attention from many students and professionals. Far too many investigators choose the response variable, the methodology for measuring it, and the evaluation unit blindly, following tradition as learned in courses or presented in the literature. Often those choices are highly inappropriate to the investigator's particular conservation concern, idea for field study, or biological surroundings. Although little damage may result in some such cases, in others the faulty inferences that are drawn may lead to faulty guidelines or faulty scientific conclusions. This step and the three that follow touch on the "natural history stories" recounted in chapter 6 and reappear in a different form in chapters 8 and 9.

In some studies there's little doubt as to the response variable that you should measure. For example, if your conservation concern involves a possible difference in the sex ratio among the caimans of two lakes, one of which experiences intense poaching pressure and the other not, the response variable you measure per caiman is simply its sex. Frequently, though, the choice is less obvious. Let's say that your concern involves the invasion of weedy plant species into the forest following selective logging (figure 4.1), and you've placed evaluation units in the replicate UL and SL response units. In each evaluation unit, will you simply count the number of seedlings of invasive plants and use that count as your response variable? Or will you take into consideration that the increased light availability in SL tracts might enhance germination of many forest natives as well—thereby rendering the raw count of invasive plants inappropriate as your measure—and instead use the *proportion* of seedlings that are of invasive species? Or will you measure the growth rate of a standard number of individual seedlings, and if so over what time interval? Or will you monitor the survival rate of individual invader seedlings to sapling size, given that many or most could die before reaching the size at which they would have significant conservation consequences? Or would you simply tally the number of species of invasive plants? Numerous other options exist. Note that each choice may lead to a qualitatively different answer to your general question and therefore to different reflections, conclusions, and management decisions.

Even the familiar question regarding selective logging and forest birds, frogs, or small mammals can be answered in a number of different ways. How many ways can you define *diversity*? I'll anticipate chapters 8 and 9 by warning you now that the alternatives include number of species (as used in this illustration), number of genera, number of families, number of feeding or nesting guilds, and the frequently (mis)used "diversity indices" that combine information on both the number of species (or families or guilds) and the number of records for each. Although the difference between these alternatives may seem academic right now, rest assured that different ways of expressing diversity may also lead to quite different answers, conclusions, reflections, and (if you haven't reflected long enough) decisions on conservation guidelines.

Step 11. Select the Evaluation Unit

Note that the concepts of both "evaluation unit" and "response variable" were brought up at the beginning of the chapter in order to introduce the designs for the selective logging study (figure 4.2). In practice, though, you shouldn't choose them until this point in the design process.

There's no rule for choosing response variables, other than they should correspond to the question you've framed. As always, use common sense, your knowledge of natural history, and a close scrutiny of your question. The same applies to choosing the evaluation unit itself. Again, in some inquiries the evaluation unit is obvious. If the question involves the sex ratio in caimans of two different lakes and the response variable is sex of the caiman, then the evaluation unit is . . . the individual caiman. If, though, the question involves some response variable that's measured per unit area that you yourself choose (as in the example of invasive plants), per unit time, or per standard number of individuals, how are you going to decide on the nature of the evaluation unit? First, read chapter 6 and ask yourself whose point of view you're going to take. Second, choose evaluation units large enough that the variation among them will make sense biologically but not so large that each encompasses much of the variability you're trying to evaluate or that each takes an inordinate amount of time to assess.

For example, in the selective-logging inquiry, you could have chosen evaluation units that were much more compact: transects of 50 m traversed in two minutes, transects of 20×2 m traversed in ten minutes, and grids of twenty live traps opened for one night only as the respective evaluation units for birds, frogs, and small mammals. You'd have gotten quantitative results, although many of these might be zero. Would differences among those observations have any biological meaning? Probably not. For example, given that capture rates of small mammals in real life tend to average between 0.5 percent and 20 percent per live trap per night, with grids of twenty traps you might trap two mammals of two different species in one evaluation unit and none in a second. This variability shouldn't even be dignified with the name "intrinsic variation." Most likely it has nothing to do with the true density and diversity of mammals in the response units involved but has only to do with the chance events that result from sampling with too small an evaluation unit. With respect to sampling small mammals at most South American sites, my own choice would be an evaluation unit with enough trap-nights to generate on average at least twenty-five or thirty different individuals captured. That's a substantial investment in traps and handling time, but lower numbers obtained with less effort and expense may be essentially meaningless. Likewise, transects for frog censuses should be long enough for at least twenty-five or thirty individuals on average to be recorded in tropical cloud forest, long enough to encounter some sixty or eighty or more individuals in lowland wet forest.

On the other hand, extra large evaluation units may lose information about the small-scale variability that's most important to your question. The consequences of choosing evaluation units that are too large for the question, or too small, have been explored particularly well for plants. Numerous studies have shown, for example, that different choices of quadrat or parcel size can lead to remarkably different results and conclusions.[8] You should be at least as worried about choosing evaluation units for animals. There are many fewer resources to guide you, though[9]—other than your common sense and knowledge of natural history.

Step 12. Select the Sampling Methodology

Let's assume that you've now selected a reasonable response variable and a reasonable evaluation unit. Is your sampling methodology—the way in which you measure each response variable in each evaluation unit—reasonable as well? Put more bluntly, are you really sampling what you think you're sampling? What you think you're sampling defines the *target population* (Manly 1992). It's the target of your conservation concern or field study and of your question. For example, here the three target populations are the respective unknown assemblages of forest birds, frogs, and small mammals that actually exist at the site. It's possible that the methodology you've chosen may end up sampling some parts of those target populations better than others. The numbers you record may reflect something quite different from the true (and unknown) nature of the underlying assemblage: the *sampled population*.

What's the target population of the small-mammal study here? The underlying fauna of small mammals at the sites where you've located the evaluation units. What's the population sampled by Sherman live traps? That subset of small mammals that tends to enter those traps most readily.[10] The tendency of mammals to enter and to spring live traps varies greatly with species, size, sex, reproductive status, social status, and age. Individual mammals may be, or become, trap shy or trap happy. Therefore, the sampled population that your records reflect may differ quite a bit from the target population.

What's the target population of the frog sampling? The forest's frog fauna. What's the sampled population? Those frogs that happen to be active between ground level and 2 m of height in the understory between the hours of 2100 and 2300. Other frog species may be active in other strata or at other hours. Worse, it's even possible that the degree of concordance between sampled population and target population varies between SL and UL sites simply because the change in vegetation structure makes it easier to see frogs of some but not all species in SL forest than in UL tracts or vice versa. So, if your data suggest that there's a logging-related change in the sampled population, can you be sure that this also reflects a logging-related change in the target population?

The same can be asked of birds. Is your bird census giving you a complete picture of the local forest avifauna? Of course not. It's well known that different bird species, and even different sexes or age classes of a given species, differ in conspicuousness and therefore in the observer's ability to detect them (Ralph and Scott 1981; Ralph, Sauer, and Droege 1995). Adult males are more detectable in breeding season than at other times. Unless you've carefully corrected for differences in detectability—a difficult task in most Latin American or Caribbean forests—your data will underrepresent the true abundance of many inconspicuous species and may miss others (for example, some nocturnal species) entirely. If you opt instead to confide in "observer-free" technology by using the traditional technique of stringing mist nets in the forest understory, have you improved the match between target and sampled populations? On the contrary. The sampled population now consists of those bird individuals that happen, for whatever reason, to be flying through that certain minuscule portion of that one stratum of a complex forest at that particular time and who don't see the net in time to avoid it. Clearly, those birds aren't likely to be an unbiased sample of the site's entire avifauna, or even of the understory avifauna alone.

It's sad but true that the only sampling methodologies quite free from bias are those for higher plants or for equally sedentary animals such as oysters. Nearly every methodology for sampling mobile animals, cryptic organisms in general, or many other features of interest to the conservation professional and field ecologist is biased to a greater or lesser extent in that the sampled population differs substantially from the target population. What to do? First, ponder the natural history of the target population. Next, read the excellent text edited by Sutherland (1996) and more detailed texts or technical papers as needed.[11] Finally, reflect on whether the biases in the sampling methodology, even if considerable, will at least be consistent among evaluation units, response units, and the different levels of the design factor. If so, go ahead cautiously. If not, worry.

For example, there's no reason to suspect that the relative trapability of different mammal species will differ markedly between UL and SL sites. Thus, any substantial differences you see in mammal density and diversity probably reflect the sorts of relative changes that are actually occurring in the underlying fauna. In contrast, there's plenty of reason to suspect that understory mist nets in SL and UL sites will sample different fractions of the target population of birds. After all, selective logging is likely to create a forest with many large light gaps. Canopy-loving birds that follow the forest-air interface will normally remain tens of meters above the ground in UL forest but in SL tracts may dip into the light gaps and be encountered near the ground. Thus, mist nets may sample a greater percentage of the forest's total avifauna in SL sites than UL sites. Apparent differences—or lack of same—in your results may be false.

In short, at the beginning of any study you should think clearly about the methodology to be used, and at the end you should reflect on the biases and artifacts that the methodology you chose may have introduced. By doing that, you'll be able to judge just how far you're entitled to apply your conclusions with a clear conscience.

Step 13. Decide Whether to Subsample

For purposes of analysis, you need to characterize each response unit with a single value. Most of the time you're most interested in an average or typical value for the response unit as a whole. On occasion the unique value you seek might instead be an index to the extent of small-scale heterogeneity that exists within the response unit. In the latter case, each response unit must include several different evaluation units so that you can assess the amount of variability among those. Even if you're seeking an average value over the response unit, though, you might wish to place several different evaluation units within it. Designs 8 and 10–12 each have only one evaluation unit per response unit. Are you willing to assume that the value you obtain from that single evaluation unit, representing a small fraction of the response unit, is the typical value you'd encounter across the entire tract (response unit)? If so, then there's no need to worry. If not, you may wish to *subsample*, or take multiple measurements (use multiple evaluation units) per response unit.

Thus, subsampling provides you with an average value among the different evaluation units that you can then use to characterize the response unit as a whole and enter into data analyses (but also see the discussion of mean and variability in chapter 5). Designs 7 (figure 4.2C), 9, and 14–16, whatever their other problems, employ subsampling. Designs 7 and 9 include six subsamples per response unit, designs 14 and 15 four (with respect to time), and design 16 two per season. If you had more than twelve total evaluation units available, you could also improve the better designs—those that have six true response units per level, such as designs 12 or 13—by additionally placing several evaluation units within each response unit. Should you do so? Again, consider the advantages and disadvantages.

Clearly, subsampling's advantage is to give an increasingly accurate picture of the characteristics of the response unit as a whole. The disadvantage is the increased time and effort involved. For example, if there's time and effort available only for twelve evaluation units, how should those be distributed among replicate response units and multiple evaluation units within response units? In the case of such strict limitations, the choice is clear: maximize the number of replicate response units and don't waste evaluation units on subsampling. In other cases, though, there's no simple answer.[12] Sometimes the evaluation unit *is* the response unit in essence, so the debate is irrelevant. Other times the evaluation unit is quite large relative to the response unit, so there's no real need to cram more than one of the former into each of the latter. If, on the other hand, the evaluation units are small relative to the space and time encompassed by the response unit, or if you have reason to suspect that the evaluation units within a given response unit are quite variable among themselves, then you might wish to invest some effort in subsampling.

Most likely, as in the selective logging example, you face an upper limit on the total number of evaluation units you can handle no matter how they're distributed. You should ensure that plenty of replicate response units exist before you engage in subsampling. If you then choose to subsample, *take care that you don't start treating the subsamples from a single response unit as if they were independent response units themselves.* Hurlbert (1984) coined the term *pseudoreplication* to describe this error. For example, relative to the selective logging question, the twelve evaluation units in designs 11 and 12 refer to twelve replicate response units (six for each level), whereas in design 9 those twelve evaluation units really involve only the two response units. If your analysis treated the twelve as if they were independent of one another, you'd commit the sin of pseudoreplication.

How can you avoid committing the sin of pseudoreplication? Either take great care that you're truly replicating with respect to the question and design factor, or else—as usual—change the question. For example, if you're stuck with design 9, once again you might simply change the question to: "Is there a significant difference in the density and abundance of amphibians, small mammals, and birds between these two cloud forest tracts, one of which happens to be UL and the other SL?" By rephrasing the question, you now find yourself with six true, replicate response units per level of the design factor—but the design factor is now "tracts," not selective logging.

Step 14. Decide on the Number of Replicate Response Units per Level of the Design Factor

In this example, we've considered designs that have either no replication (i.e., one response unit per level of the design factor) or six replicate response units per level. I chose those numbers arbitrarily, for illustrative purposes only. Are even six replicate units enough? Probably not, but then how many *are* enough and how do you decide that? Chapter 5 and appendix B will deal with this fundamental concern.

Step 15. Decide on the Time Frame Available and Make Sure It's Adequate

Parts of chapter 6 will address this point, and see Crome (1997). Clearly, any effects of selective logging that you record during a field experiment, or today's observed differences between already logged and UL sites, are not fixed values. Those effects and differences change as the forest regenerates. An experiment, such as design 13, will probably reveal an immediate, strong effect of selective logging on any response variable you care to examine, but as the forest recovers, might not those short-term responses have grossly overestimated the long-term effects most significant to conservation goals? If you don't reflect on this point, you might find yourself proposing unnecessarily strict management guidelines. Conversely, the cumulative, long-term effects of mercury contamination from clandestine gold-mining activities within your reserve may be much more pronounced than the effects you can measure in a short-term study. Unless your study can last for several years, you may grossly underestimate the real consequences.

If, following reflection, you must honestly answer no to the question "Is the time adequate?" be extra cautious in your interpretations and applications.

Step 16. Decide on the Amount of Effort Available and How to Allot It

Refer to steps 2, 3, 7–12, and especially 13–15 above, then set up the precise layout of the study in time and space, specifying not only the response units but also the evaluation units. If you don't go through this preliminary step, you might find that you've personally committed thirty-two-plus days per month to the study and still don't have adequate replication.

Step 17. Decide Provisionally on the Way You'll Analyze and Present the Results

Steps 2, 4–6, 9–12, and 14 touch on these decisions. Chapter 5 goes into much greater depth on statistical analysis but is by no means sufficient. If after reading chapter 5 you decide that statistical inference will be valuable, consult one or more of the texts cited there, and better yet consult a statistician, before going to the next step.

Step 18. Get to Work!

But first you might wish to practice. Have a look at box 4.2.

Recognizing Your Limitations

If you succeed in completing a well-designed study within the scope of the question, you can feel reasonably confident (see the next chapter) in your conclusions regarding the design factor. After reflecting long and hard, you can use this information as one crucial element in formulating, proposing, and implementing conservation guidelines, or in suggesting new lines of research brought up by your interesting results. Perhaps, though, upon reading through the eighteen-step list above, you're despairing of ever being able to complete a well-designed study. Does the design process seem too complex? Too costly? Too confusing? Just plain overwhelming? Should you just forget it and set

---

Box 4.2. Practice with Design

Select one of the questions you thought up during the exercise of box 3.2. How would you design a study to answer it? Alone or with colleagues, practice following the steps in this chapter to design a complete study. Then ask yourself if you feel capable of designing real-life studies related to your own most pressing conservation concerns. If the answer is no, try practicing with another of your questions from box 3.2.

up guidelines based on your gut feelings or based on what other people and other studies tell you? Once again, no. Do the best you can, in your own landscape. Just reflect long and hard before deciding how, or whether, to apply the results. Of course, the results of your study, or any other application of scientific inquiry to management of protected areas, won't make the decision for you. That's up to you. The inquiry is only one of innumerable social, political, legal, ethical, biological, and pragmatic factors that you must take into account.

Rest assured that there's no such thing as a perfect study. Indeed, trying to achieve that perfection may involve some serious ethical drawbacks (box 4.3). Managers, conservation ecologists, and other professionals must recognize that their results, like those of any other scientific inquirer, only apply within certain limits. Scrutinize in detail the published studies on which you're tempted to base conservation decisions, or papers in your area of field ecology, and you'll find that every one falls into one of the following categories:

1. With respect to the scope of the question, there was a serious design flaw (often, pseudoreplication) apparently unrecognized by the author. The analysis and interpretation of the results are suspect.

2. There was a serious design flaw surely recognized by the author, but the author succumbed to the common temptation to "explain the world," squelched his or her doubts, and brazenly extended the conclusions far beyond their reasonable limits.

3. The question involved events taking place at such a large scale in space and time that replication was minimal. In theory, then, there was a serious design flaw. Nevertheless, the author, being a cautious scientist and good natural historian, argued reasonably that the great difference between results for different levels of the design factor, differences orders of magnitude greater than those expected from possible confounding factors or intrinsic variation, could be explained only by the design factor itself.[13]

4. Despite replication several possible confounding factors existed, some of them possibly quite severe. The author, though, took pains to phrase the question with this in mind and to reflect thoroughly before drawing cautious, well-reasoned conclusions.

5. The study was beautifully designed but for such a narrow question and over such a narrow scope of conditions that the conclusions could scarcely be extended to any other time or place.

Monitoring

Of course, even if there's no such thing as a perfect study at least you have the choice of designing one that's less imperfect or more imperfect. You undoubtedly encounter the word *monitoring* just about as often as you encounter the word *biodiversity*. Just what is monitoring? Monitoring is in the eye of the beholder. I know of no one definition that's accepted universally.[14] Nevertheless, three alternatives cover most cases:

1. *Monitoring is keeping track of what's happening by recording data continuously over time.* If you sample well over the entire scope in question, and if samples are collected in objective and consistent fashion, then monitoring by this definition provides a sound database. Following reflection, these data may generate questions complying with the guidelines of chapter 3, may be used as a baseline in other studies, or may be summarized and examined for trends over time.

Any apparent trends in the data cannot be further interpreted, though, because at present no basis for comparison exists.

2. *Monitoring is keeping track of what's happening with the goal of keeping an eye out for "surprises" or suspicious trends.* The only way to recognize a surprise or a suspicious trend is to have baseline or control data for comparison. Data that diverge "significantly" from the baseline or control data should be cause for worry. It's important to recognize, though, that this method is not capable of objectively assigning the blame for any surprises or suspicious trends that do occur.

3. *Monitoring is an investigation that follows all the rules for first-hand inquiry* and is often oriented toward the question "What's the effect of human intervention in this system?" or "What's the effect of implementing such and such a management guideline?" The selective logging question as a whole resembles the former question and thus might qualify as monitoring, if the units are sampled repeatedly over a long time period. The explicit "monitoring" step in the management cycle (figure 2.4) is in essence the latter question. By this definition, then, monitoring should be designed with all the rigor of any inquiry and should pass through the eighteen steps above.

Which definition fits your current monitoring program? Given that there's a great difference

---

Box 4.3. An Ethical Dilemma

The stronger the study design, the greater your confidence in conservation guidelines that are based on the conclusions (see figure 1.1). The way to accomplish a strong design is to replicate extensively across the landscape in question, and sometimes to subsample as well. Every evaluation unit you install, however, molests to some extent the site, flora, and fauna involved. In the selective logging example, clearing paths for the bird and frog transects disturbs the forest understory and its biota to some extent, and walking the transects compacts the soil. Despite the name, live-trapping will kill or injure some small mammals. Furthermore, as pointed out, the only way to truly study the effects of the design factor is to carry out an experiment. Experiments at the scale of most management concerns are highly intrusive, clearly disrupting the lives of many plants and animals. Witness the logging experiment proposed in design 13. In short, your dilemma is that the best-designed studies are the most disruptive, while the gentlest studies will be the weakest in terms of their applicability to guidelines whose purpose is conservation of the protected area.

To what extent, then, can you justify disrupting the landscape and its living inhabitants in order to provide guidelines for conserving them? Must you, to quote a U.S. military officer during the Vietnam War, "destroy the village in order to save it"? As always in this primer, the answer is up to you. In conservation or in basic field studies, the question of ethics is not black and white (Farnsworth and Rosovsky 1993). Just ponder profoundly the lead question of this paragraph—and the quote from the Vietnam War—as you design your study. Perhaps another compromise on design is in order.

among the three definitions in the sorts of conclusions you can draw, and decisions you can later make, which definition would you *like* to fit your current monitoring program?

Conclusion—For Now

Design is a process of compromise. You must compromise between the ideal design and the real limitations of time, effort, finances, and landscape. You must compromise between doing the best you possibly could within those limitations and minimizing the negative impacts the inquiry has on the system you're trying to conserve. Maintain a balanced perspective throughout the process, and your design will indeed allow you to proceed through the management cycle or the field research cycle—once you've absorbed the fundamental concepts that chapter 5 presents.

CHAPTER 5

Small Samples and Big Questions: The Role of Statistical Inference

NEVER confuse statistical significance with biological significance.

—Charles J. Krebs (1989)

Most versions of the logging question that dominates chapter 4 contain the words *significant* and *significantly*. What's meant by those words? Sometimes we mean the synonyms that you'll find in a dictionary—for example, *important, notable, serious, having meaning.* If we're dealing with some aspect of living organisms, we're talking about *biological significance.* Often, though, we intend a quite different definition, *statistical significance.* As Krebs (1989) warns, those last two meanings of significance are often confused, and that confusion can have serious consequences for the interpretation of any inquiry, whether the question involves basic science or conservation.

This chapter will attempt to sort out the confusion. It presents the fundamentals of some powerful, practical tools for proceeding far beyond your design and the set of results, to the point where you may be able to draw inferences and make cautious decisions about the much larger scale in space and time that your sample is meant to reflect. By chapter's end it should be clear that statistical significance and biological significance play lead roles, but very different ones, in this process. You'll see that the development of realistic conservation strategies (Margoluis and Salafsky 1998; The Nature Conservancy 2000) must include the philosophy, if not always the practice, of *statistical inference,* and you'll see that the reflection phase of any inquiry at all can scarcely take place without taking that philosophy into account. After all, the patterns that exist in your data set could easily have arisen by chance. You can't blame those patterns on the design factor unless you can separate the different sources of variation. If used properly, statistical inference is the ideal—or the only—tool for that purpose and for many others as well.

To follow the reasoning of this chapter you need not have any prior experience with statistics. Indeed, you might grasp the philosophy and logic better if you haven't. I'll begin with the basics, the ways to present the average value and the extent of variation among your data, before embarking on a critical, step-by-step presentation of the philosophical and logical basis for statistical inference. Near the end of the chapter I'll briefly mention some sophisticated approaches currently used by some conservation professionals and field ecologists. Up until those final paragraphs, despite recent controversy (e.g., Crome 1997; Johnson 1999) I'll emphasize "classical" statistical inference. Those of you with any statistical training will be most familiar with that (as I am), and, for better or worse, it's still by far the most widely used approach to basic and applied inquiries.

Although you need no prior experience with statistics, classical or otherwise, once again I'll need to bring up a fair amount of terminology (box 5.1). This time the terminology includes a number

---

Box 5.1. An Informal Glossary for Chapter 5

Arithmetic mean: Often labeled just *mean,* the average value over the set of observations (compare with "Median" below). Sample mean is $\bar{x}$, population mean is μ.

Biologically significant: Meaningful from the point of view of the organisms, ecological processes, landscapes, or conservation concerns involved. For example, in a well-designed study, an average reduction in number of frog species from twenty to five following selective logging is certainly biologically significant, whereas an average reduction from twenty to nineteen is almost certainly not, regardless of what the statistical test tells you. Contrast with "Statistically significant" below.

Classes of data: The different ways to characterize the nature of the response variable. Don't confuse these with the different kinds of levels of the design factor (box 4.1). *Interval data* are those obtained when the response variable is measured in such a way that (in theory) the interval between any two measurements could take on any value, including a fractional one. *Ordinal data* are those obtained when different observations can be ordered with respect to one another or can be assigned ranks. Often, two or more observations can receive the same rank. *Nominal* or *classificatory data* are those obtained when different observations can be assigned to distinct categories but those categories can't logically be ranked greater or lesser with respect to one another, as in categories of sex, species, or eye color.

Coefficient of variation (CV): A measure of variability that's standardized in relation to the mean so that the relative amount of variation in two or more samples can be compared.

Confidence interval: The interval of values, defined by the two *confidence limits,* within which you suspect the unknown true value of a population statistic (parameter) really lies. This is a practical, working definition only. The exact meaning is as follows: If you took an infinite number of samples of size n from the statistical population and computed the XX percent confidence interval each time, XX percent of those calculations would include the true value of the parameter.

Degrees of freedom (df): A whole number (i.e., ≥ 1) that's often related to sample size n and is always defined by the particular statistical procedure you're using. Practically speaking, you need be aware of, or use, this value only when you're referring to the statistical table for that procedure.

Effect size (e): The true but unknown magnitude of the effect of the design factor on values of the response variable in the relevant statistical population(s). For inquiries with categorical levels of the design factor and a response variable with interval data, this would be the *difference (δ)* between the mean values for the different levels.

Median: The value of the observation that's exactly halfway between that with the lowest value and that with the greatest value. That is, it's the observation below which 50 percent of the other observations lie and above which the other 50 percent lie. Compare with "Arithmetic mean."

Meta-analysis: A method for combining the statistical results of different studies, done at different places and times and by different investigators, in order to estimate a universal, average effect size (e) of a given design factor.

Multivariate statistics: The large number of techniques for dealing with data sets that have several response variables that may be correlated with one another, such as counts for different tree species in a series of vegetation samples.

Observation: Also called *datum* (plural *data*), the response variable's particular value or characteristic (x_i) that's associated with a given response unit or evaluation unit i.

Parameter: A single measure that summarizes one kind of quantitative information about the statistical population as a whole. You can almost never calculate its precise value, but statistical inference allows you to estimate that. Parameters, occasionally termed *population statistics,* are often denoted by Greek letters. Contrast with "statistic" below.

Power analysis: The technique used to estimate the power of a statistical test before you actually collect the data (*a priori* power analysis) or to try to figure out the power of a statistical test that you've just performed on data already collected (*a posteriori* power analysis). At least in our field, the second use is not considered proper.

Power of a statistical test: The probability that your particular statistical test will be able to detect a true effect size e (or δ, where the design factor has categorical levels) or a larger one if e (or δ) exists in the statistical population(s). That is, it's the chance that you'll correctly reject the null hypothesis if that e (or δ) truly exists. *Statistical power* is the same idea in more inclusive terms, sometimes referring to comparisons between different kinds of statistical tests—for example, parametric vs. nonparametric approaches applied to the same set of data.

Rejection level: The maximum risk you'll accept of committing a Type I statistical error should you decide to reject the null hypothesis, or α_{rej}. In some texts this is called simply the *alpha level* or α *level.* You decide on α_{rej} before beginning the study, and after the study compare that with the α resulting from the data, or α_{obs}—also known as the *P*-value. If α_{obs} exceeds α_{rej}, you cannot reject the null hypothesis under the conditions you've set. If α_{obs} is less than α_{rej}, you can reject the null hypothesis as long as you remain aware that there's always a chance of being wrong, of magnitude α_{obs}.

Resampling statistics: Computer-intensive methods of statistical inference in which estimates of the parameters of the statistical population, or evaluations of the statistical null hypothe-

Box 5.1. Continued

sis, are generated by the sample itself. The computer reshuffles the original observations a very large number of times, following the philosophy and technique of the particular approach. Sometimes the approaches used to evaluate statistical null hypotheses per se are called *null-model statistics.*

Sample: The set of data or observations you record during the study. *Sample size,* or *n,* is the total number of such observations unless, as in appendix B, it refers to the particular number per categorical level of the design factor.

Standard deviation: One measure of the variability among data, the square root of variance. Sample standard deviation is *s,* population standard deviation is σ.

Standard error of the mean or $SE_{\bar{x}}$: As calculated for a sample of size *n* and mean $\bar{x}$, an estimate of the standard deviation that you'd record among a large number of such means if you were to continue to draw samples, all of size *n,* from the same statistical population.

Statistic: A single measure that summarizes the information in a set of data (sample). Sample statistics are labeled with Roman letters and are often used to estimate the values of the unknown parameters (see "Parameter" above).

Statistical error: A mistake made when you decide to reject, or not to reject, the null hypothesis. A *Type I error* is deciding to reject the null hypothesis when you really shouldn't; the risk of so doing is α. If used correctly, a given statistical test will provide you with the precise estimate of that risk, α_{obs}, sometimes called *P* or the *P*-value. A *Type II error* is deciding not to reject the null hypothesis when you really should; the risk of doing so is β.

Statistical hypotheses: The pair of statements about the underlying relationship in the statistical population between the design factor(s) and the response variable. Usually the *null hypothesis* (H_0) states that there is no relationship, while the *alternate hypothesis* (H_A) states that there is.

Statistical inference: The process of making educated guesses about the underlying quantities or quantitative relationships that your sample presumably reflects. *Classical statistical inference* treats those unknown, underlying phenomena as fixed and your data as a random sample. Two classical approaches exist: You're doing *parametric statistical inference* if your educated guesses concern the parameters, most often the mean and variance of one or more statistical populations. *Nonparametric statistical inference* depends not on those particular parameters but on other features of the data. In contrast to classical techniques, *Bayesian statistical inference,* mentioned only briefly here, treats your data as fixed and proposes values for the underlying phenomena based on those data.

Statistically significant: In the "α-heavy" statistical inference traditionally used by workers in our field, having an α_{obs} (*P*-value) that's equal to or lower than the α_{rej} that was chosen before the study, most often 0.05. If α_{obs} is especially low, say with a value of 0.01 or 0.001, the results are often termed "highly significant." Both designations are highly subjective. The latter simply means that there's very little risk of committing a Type I error should you decide to reject the null hypothesis. Whether these labels have any relation whatsoever to biological significance is quite another question.

Statistical population: The universe of all possible observations about which you wish to draw conclusions based on your sample of a limited number of such observations.

Statistical test: Any one of a great number of methods for extracting a value of α_{obs} (the *P*-value) from a set of data collected with respect to a comparative question. This value is the risk of making a Type I statistical error should you decide to reject the null hypothesis for the statistical population(s) as a whole.

Tabulated value: For a given set of data subjected to a statistical test, the value, as found in a table for that particular class of test, for the test statistic that corresponds to the particular α_{obs} and degrees of freedom. Alternatively, the tabulated value for α_{obs} is that which corresponds to the observed test statistic and the degrees of freedom.

Test statistic: The value that results from applying a particular class of statistical test to your particular data set. You should always report this value, along with n, α_{obs}, the degrees of freedom, and the sample statistics themselves. With the value for the test statistic, you turn to the table for that class of test and locate the value of α_{obs} for that data set.

Variance: An extremely useful, if nonintuitive, measure of variability among data: the average of the squared differences between the value of each of the observations and the mean value. Sample variance is s^2, population variance is σ^2.

of symbols, which are often Greek letters. These symbols find their way into a number of mathematical equations. If equations and Greek letters inspire fear and trembling, consider the following. First, these equations are just shortcuts, and they involve simple arithmetic. Second, they're not essential to an understanding of the underlying philosophy as long as you read the surrounding text carefully. Skip over them if they bother you a lot. If you still don't find the chapter to be user friendly, skip all the way to appendix D; then return to this chapter better armed.

Why Bother with Statistical Inference at All?

Simply by understanding the philosophy and basic approach of statistics, you'll view the inferences and conclusions that spring up during reflection in a new light, you'll reevaluate the decision-making process that follows, and you'll ponder the risks and the consequences to conservation of making the wrong decision. There are as well more pragmatic, if somewhat cynical, reasons to learn about statistical inference, especially statistical tests (see Johnson 1999). First, statistical inference gives the impression of being an objective and exact means of spanning the gap between your small data set and your big question, even though it doesn't always live up to this. Second, everyone else seems to use statistical inference—how many times have you run across "significant," confidence limits, results of statistical tests, *P*-values? Third, many computer packages of statistical techniques are available; it's very easy to toss your data into a program and get some awe-inspiring, if often misunderstood or irrelevant, numbers out the other end. Fourth, students, field biologists, and conservation professionals are taught to use statistics and are castigated if they don't. Fifth, because they themselves went through that experience, many journal editors, thesis supervisors, and heads of conservation projects demand statistical analysis of any data set. If you yourself understand the philosophical basis and proper use of statistical inference, you'll be able to soundly criticize the all too frequent cases in which it's abused, misused, and misinterpreted (Yoccoz 1991; Johnson 1999). You'll also be better able to defend your choices—for example, the choice *not* to apply a statistical test—successfully against those editors, professors, and project heads.

If the fundamentals presented in this chapter aren't enough, if you truly yearn for mathematical models and cookbook recipes, you can turn to literally hundreds of texts[1] and dozens of computer packages.[2] The intent here, though, is that before doing so you will recognize how to use properly the potentially dangerous tools of statistical inference. If you have no intention of using those tools, if you've never before looked a *P*-value in the eye or intend to do so in the future, I hope you'll at least apply the philosophy of statistical inference to what you do or what you review. On the other hand, even if you're already a statistics whiz, you may find some surprises in what follows—and you'll certainly find points with which to disagree, for many of the issues discussed below are subjects of heated controversy among statisticians themselves.

Descriptive Statistics

In general usage, a *statistic* is simply a single value that summarizes the information contained in a set of observations or data. Often (although not always) we're most interested in a pair of statistics that summarize the information in quite different ways: one that presents the typical or average value among the observations, another that presents the extent of variation among them. Often the statistic chosen to represent the average value is the *arithmetic mean* (from now on called simply the "mean"), which is the sum of the individual observations, each called x_i, divided by the total number of such observations involved or *sample size, n*:

$$\text{mean} = \frac{\sum_{i=1}^{n} x_i}{n} \tag{5.1}$$

where $\sum_{i=1}^{n} x_i$ is the command to sum all the different instances of what follows, in this case the individual observations (x_i) from the first $(i = 1)$ to the last or n^{th}. Often the statistic chosen to represent variation among observations is *variance,* the average of the squared differences between the individual observations and the mean. The calculation of variance is a bit more complex but still easily accomplished by hand:

$$\text{variance} = \frac{\sum_{i=1}^{n} \left[(x_i - \text{mean})^2 \right]}{n} = \frac{\sum_{i=1}^{n} x_i^2 - \left[\frac{\left(\sum_{i=1}^{n} x_i \right)^2}{n} \right]}{n} \tag{5.2}$$

The left-hand formula is easier to understand; the right-hand formula is easier to use if you're employing a hand calculator. Many times variation is also expressed as *standard deviation,* the square root of variance:

$$\text{standard deviation} = \sqrt{\text{variance}} \tag{5.3}$$

For example, figure 5.1 shows two data sets with the same mean but different variances. The observations in set A display greater variability, and thus will display larger values of variance and standard deviation, than those in set B.

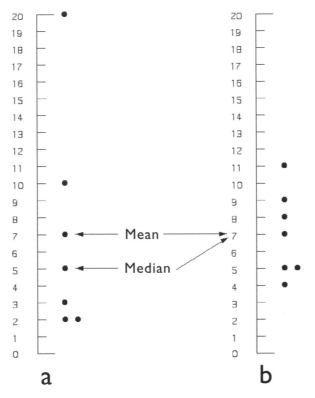

Figure 5.1.
Variability. Two data sets with the same mean (7) but different amounts of variation.
Each dot represents a single observation. Note that in A the median is quite different from the mean.
Which one would you choose as the statistic that best represents the average value?

Most discussions below will involve the two statistics of arithmetic mean and variance, although alternatives exist and are discussed below. For example, let's say you've subsampled within response units (as in step 13 of chapter 4), but now you must characterize each response unit by its own, single value for the response variable of interest (see below). You could obtain that overall value by calculating one or the other statistic across the individual values from the subsamples (different evaluation units). More commonly, you wish to compare those values, one per response variable, among the various response units themselves that correspond to the different levels of the design factor. You can do so by computing mean and variance among the response units themselves, not the subsamples, for each level.

Some Important Alternatives to Mean and Variance

Mean and variance (or standard deviation) aren't the only statistics available for expressing the average value and the amount of variability, respectively. The leading alternative to the mean is the *median,* the value of the observation that's exactly halfway up the list from the smallest to the largest. Thus, in a set of seven observations (e.g., figure 5.1), the median is simply the value of the fourth smallest (or largest) observation, whereas in a set of eight observations, the median's value is exactly halfway between the fourth and fifth observed values—i.e., the average of those two data.

Does it matter to your inquiry whether you choose median or mean? Sometimes it matters a lot. There's a fundamental difference in what the two measures represent. *If you need the average value among observations, choose the mean; if you need the value of the typical observation, choose the median.* For example, let's say that you're concerned about an endangered understory tree species whose fruits are consumed, and seeds dispersed, by one bat species. Over a large number of trees, you record the number of fruits removed per night, presumably by bats. The results are highly variable: on the great majority of trees only a few fruits or none whatsoever are removed, whereas a very few trees enjoy astoundingly high rates of fruit removal. Because of the numerical influence of those few popular trees, the mean and the median you calculate differ greatly from one another, much more than those in figure 5.1a. The value of the mean is much higher than that of the median. What does the mean value tell you? *The average number of fruits removed per tree.* What does the median value tell you? *The number of fruits removed from the typical tree.* Whose point of view does the mean better represent? That of the bats and their overall activity: for instance, it gives us an index to the total amount of fruit consumed by the bat population. Whose point of view does the median better represent? That of the trees: it suggests that the typical tree in the population actually disperses very few seeds, at least this year.

The choice between mean and median may especially matter if you're subsampling within response units and using the average value of the response variable among the subsamples in order to characterize each response unit as a whole. If the subsamples within a given response unit display many small values and a few big ones, such as in figure 5.1a or a case much more extreme, the mean and median may differ so much that your choice between the two greatly affects the outcome of your analysis. Which should you choose? That depends on your question and on whose point of view you're taking.

Alternative statistics to represent variability also exist. For example, if you're using the median to express the average value, then you might express variability through the values of the first and third *quartiles.* The first quartile is that value below which 25 percent of the observations lie and above which the other 75 percent lie; the third quartile is that value below which 75 percent of the observations lie and above which the remaining 25 percent lie (see Zar 1999). The greater the variability among the observations, the farther apart will be the first and third quartiles. Even if you express the average value with the mean, which is most common, you have different choices for representing variability. Variance and standard deviation express the absolute amount of variation. In contrast, the *coefficient of variation,* or *CV,* expresses the amount of variation relative to the mean:

$$CV = (\text{standard deviation})/(\text{mean}) \qquad\qquad (5.4)$$

If you're comparing the extent of variability in two or more data sets, should you choose *CV* or standard deviation (or variance)? If the means of the data sets are similar, it doesn't matter. The result will be the same or nearly so in either case. If the means are quite divergent, though, your choice does matter. Imagine that you're comparing the variability in, not the average of, fruit production by the bat-dispersed understory tree species mentioned above in selectively logged and unlogged sites (see figure 4.1). The understory trees in SL sites receive much more sunlight, thanks to the thinning of the canopy over them, and on average produce twenty times as many fruits as trees at the UL sites. Simply because all the values are higher, standard deviation or variance (i.e., an absolute measure of variation) in fruit production among trees at SL sites almost certainly exceeds that among their UL counterparts. You discover that the reverse occurs, though, when you correct for the strong effect of that difference in means, by calculating coefficients of variation. The *CV* for trees in SL sites is actually much smaller than that among trees at UL sites. That is, *relative to the mean*, the trees at SL sites produce fruit more consistently than do trees at UL sites.

So again, which statistic should you choose? If taking the bats' point of view, choose the absolute statistic, standard deviation. This measures the variability (or unpredictability, to the bat) in real numbers of fruits, or in absolute availability of food, from tree to tree. If taking the trees' point of view, though, consider choosing coefficient of variation. This indicates that relative to the resources available to them, fruit production varies more among the trees in the UL sites than among those in the SL sites. This implies that the contributions of different trees to the local gene pool are likely to be more variable among trees in UL sites than among those in SL sites. If conserving the genetic diversity of this particular tree species is an urgent management concern, then you might actually consider encouraging an increase in selective logging in order to even out the relative contributions of the different trees. I trust that this wouldn't be the only criterion on which you'd base such a decision, though.

Another way of comparing the relative variability of two samples is to convert each observation to its logarithm and calculate the variance (or standard deviation) among those logarithms rather than among the original values themselves. This simple procedure corrects for the tendency for absolute variability among the data to increase as the mean increases. Nonparametric methods for comparing variability among samples (see below) also exist. Please note that it's possible that none of these measures truly measures the sorts of variability that are most important to living organisms or ecological processes. It could be that the occasional, extreme values of the response variable are far more important than the "typical" variability of that measure. Are there ways of characterizing the variability of a data set in terms of its most extreme values instead of its "typical" variation? Yes. Gaines and Denny (1993) propose an excellent technique. But don't try their method unless you're a math whiz.

Do You Mean Mean, or Do You Mean Variability?

If you've subsampled within each response unit, you will need to characterize the response unit with a single value before proceeding with the data analysis. What's the better choice for characterizing the response unit as a whole: the average among the subsamples or the extent of variability among them? As always, the answer depends on your question. We've all been taught to think in terms of average or typical values, so without further question most workers select the mean, or another statistic for the average, as the value with which to characterize the response unit. With respect to some management concerns and questions, though, a measure of variability makes more sense. What's the biological interpretation of a statistic of variability among subsamples? The extent of heterogeneity, or unpredictability, among the different evaluation units within the response unit. This may be just what you want to know. After all, many forms of human intervention alter the degree of heterogeneity, or conversely the predictability, of the landscape's biological and ecological features.

For example, your question might be whether selective logging alters the degree of small-scale spatial heterogeneity in the abundance and diversity of small mammals. To answer that, you'd scatter a number of evaluation units over each of the replicate response units for each level of the design factor, those being SL parcels and UL tracts of about the same size. You could calculate variance, standard deviation, coefficient of variation, or the difference (third quartile minus first quartile) among the various subsamples per response unit and then compare the resulting values from all the replicate SL parcels with those from all the replicate UL tracts.

When framing your question and choosing your response variable, you must ask yourself what's more important to the management concern, changes in the average value of the response variable or changes in its variability. Only then should you select the corresponding statistic—or you might decide to ask about, analyze, and interpret both. In short, even the very simplest means of statistical analysis—summarizing the data—involve choices that require some hard thinking about what's significant biologically. If at the outset you recognize the biological meanings and consequences of different choices, you'll probably select those that make the most sense.

Statistical Inference: Introduction

If you were able to obtain every possible observation within the universe that's relevant to the scope of your question, you'd have recorded the entire *statistical population* of values (not to be confused with a biological population except under very special circumstances, as in the example below). The statistics calculated would then be called *parameters* and would be designated by Greek letters. The *arithmetic mean of the population* is designated μ and calculated with equation 5.1 in the previous section. The *population variance* is designated σ^2 and calculated with equation 5.2, while as usual the *population standard deviation*, or σ, is simply the square root of σ^2.

Almost always, though, your data represent only a *sample* from the universe of possible observations relevant to the scope of your question, that is, the statistical population of interest. And almost always you wish to extrapolate from that limited sample to the larger scale for which you'll propose guidelines (see figure 2.4) or draw conclusions (see figure 2.5). Refer to design 11 in chapter 4 for the selective logging question. If you were interested only in those twelve tracts (and only the particular evaluation units within each) at only those particular moments in time, not in anything else, your data set would be sufficient. There'd be no need to draw inferences about any larger universe that these data might reflect. The selective logging question refers, though, to all times and places where SL might take place within the reserve's perimeter. The six SL tracts you've examined represent only a sample of all possible SL tracts within the reserve (figure 4.1), the UL sites only a sample of a very large number of possible UL forest areas. Likewise, if you've carried out a manipulative study such as in design 13, your experimental results represent only a sample from the infinitely large statistical population of the results you'd obtain if you were able to repeat the experiment time after time. You don't know whether your sample perfectly mirrors or badly misrepresents the statistical population of all possible results from which it was drawn. So how can you exploit what's known, the sample, to say something about what's unknown, the statistical population? Through statistical inference.

Classical statistical inference is based on mathematical theory about the relation between a sample of *n* data (observations) and the set of all data that make up the universe of interest: the statistical population. In other words, statistical inference concerns your particular set of data and those data you'd obtain if you were able to continue sampling, or experimenting, under exactly the same conditions. Often statistical inference revolves around the complementary parameters μ and σ^2 and is thus termed *parametric statistical inference*.[3] Note that statistical inference of any variety almost always requires that (1) you've sampled randomly from the statistical population of values that pertain to your question and (2) each observation is entirely independent of the others. Many (not all) forms of parametric statistical inference also assume that (3) the values in the statistical population display a *normal distribution*. Some approaches require other assumptions as well. A discussion of the normal distribution and all that's derived from it is well beyond the scope of this manual. Most statistics texts, such as those in note 1 for this chapter, treat the subject in depth.

Of course, the only statistics you can calculate in practice involve your sample of size n, not the entire statistical population. Thus, you're actually working with *sample statistics*, not with parameters. The *sample mean*, designated $\bar{x}$, is calculated just as in equation 5.1, or:

$$\bar{x} = \frac{\sum\limits_{i=1}^{n} x_i}{n} \tag{5.5}$$

The equation for *sample variance* or s^2, though, differs slightly from that for population variance—keep your eye on the denominator:

$$s^2 = \frac{\sum\limits_{i=1}^{n}\left[(x_i - \bar{x})^2\right]}{n-1} = \frac{\sum\limits_{i=1}^{n} x_i^2 - \left[\frac{\left(\sum\limits_{i=1}^{n} x_i\right)^2}{n}\right]}{n-1} \tag{5.6}$$

Sample standard deviation, or s, is, as usual,

$$s = \sqrt{s^2} \tag{5.7}$$

On the one hand, by summarizing the set of n observations, the sample statistics describe the complete sample itself, just as in the previous section. On the other hand, $\bar{x}$, s, and s^2 can also be considered as estimates of the unknown parameters μ, σ, and σ^2, respectively, of the statistical population from which these observations are drawn. How good are sample statistics as estimates of population statistics? Since you're more likely to be interested in those estimates than just in the values of the sample statistics for your restricted data set, let's find out by asking a simple noncomparative question and playing with some numbers.

Estimating What You Can't Know Directly

Doctora Felícite Navideña manages Parque Nacional El Perdido, in which Laguna Olvidada supports a biological population of the globally endangered ferocious caiman, *Neosuchus enojadisimus*. To report to her superiors on the status of the population, she must estimate the average size of individuals. So her simple, noncomparative question is "What is the mean length among adult and subadult caimans in this lake?" Let's say that you and I are omniscient (during this exercise, that is), while Dra. Navideña is not. You and I know that there are precisely ninety-nine adult and subadult caimans in the lake, and we know the length of every one (table 5.1). By applying equations 5.1–5.3, respectively, we even know the true parameters: $\mu = 2.2$ m, $\sigma^2 = 0.79$, and $\sigma = 0.89$.[4] Dra. Navideña, though, has no idea of how many caimans the lake holds, or of the true parameters of their lengths. All she can do is sample. Let's see how well her sample statistics estimate the parameters.

First, Dra. Navideña captures and measures a single caiman ($n = 1$). Does this provide her with an estimate of μ? Yes. After all, the caiman in question comes from the statistical population in question, it contributes to the unknown value of μ, and therefore it's an estimate of μ. If there's the slightest intrinsic variation among observations, though, by chance a single observation may be way off as an estimate, and the investigator has no way of knowing that. For example, the caiman that Dra. Navideña samples at random is #89 (table 5.1),[5] and we, being omniscient, know that the value

Table 5.1. Lengths in meters of all adults and subadults, numbered from 1 to 99 for purposes of the exercises in the text, of a biological population (and, in this special case, statistical population) of the ferocious caiman, *Neosuchus enojadisimus,* living in Laguna Olvidada, Parque Nacional El Perdido, Lemuria.

1: 1.7 m	21: 2.1 m	41: 1.9 m	61: 6.1 m	81: 1.9 m
2: 3.2 m	22: 3.1 m	42: 2.7 m	62: 1.7 m	82: 2.4 m
3: 0.9 m	23: 1.7 m	43: 1.6 m	63: 2.4 m	83: 1.0 m
4: 2.7 m	24: 1.4 m	44: 2.4 m	64: 2.2 m	84: 2.7 m
5: 2.4 m	25: 2.6 m	45: 4.1 m	65: 1.9 m	85: 2.0 m
6: 1.9 m	26: 2.9 m	46: 2.2 m	66: 3.0 m	86: 1.9 m
7: 0.8 m	27: 2.3 m	47: 2.5 m	67: 2.3 m	87: 0.6 m
8: 2.5 m	28: 0.9 m	48: 2.4 m	68: 1.4 m	88: 2.6 m
9: 2.3 m	29: 2.9 m	49: 2.2 m	69: 2.0 m	89: 4.0 m
10: 4.3 m	30: 1.9 m	50: 1.3 m	70: 3.1 m	90: 2.4 m
11: 2.7 m	31: 2.2 m	51: 3.1 m	71: 2.9 m	91: 1.3 m
12: 2.6 m	32: 2.4 m	52: 1.9 m	72: 1.7 m	92: 1.7 m
13: 1.8 m	33: 3.8 m	53: 2.0 m	73: 2.6 m	93: 2.6 m
14: 2.9 m	34: 5.1 m	54: 1.7 m	74: 1.4 m	94: 2.8 m
15: 1.7 m	35: 2.1 m	55: 3.0 m	75: 2.7 m	95: 0.9 m
16: 2.4 m	36: 1.6 m	56: 1.6 m	76: 1.3 m	96: 1.6 m
17: 2.7 m	37: 1.3 m	57: 1.0 m	77: 2.0 m	97: 1.9 m
18: 0.7 m	38: 2.7 m	58: 3.3 m	78: 2.5 m	98: 1.4 m
19: 2.4 m	39: 2.4 m	59: 2.4 m	79: 0.7 m	99: 2.8 m
20: 1.3 m	40: 1.8 m	60: 1.8 m	80: 3.1 m	

of 4.0 m is quite poor as an estimate of μ. Does the quality of her estimate improve with a larger sample size? Let's say she grabs caimans 55 and 39 as well as caiman 89, for an n of 3 and an $\bar{x}$ of 3.2 m—which you and I know to be a better estimate, although it's still not great. Also, she can now calculate s^2 for this sample, applying equation 5.6 to the three values 4.0, 3.0, and 2.4 m. The s^2 of 0.65 that results is also an estimate. It estimates σ^2, that is, variation within the statistical population—again, not very well in this case. Dra. Navideña then increases n to 10 by sampling at random caimans 22, 51, 78, 2, 56, 8, and 28 in addition to the first three, yielding an $\bar{x}$ of 2.6 m (which we know to be still too high but not so high as the previous estimate of μ) and an s^2 of 0.77 (by chance a very good estimate of σ^2). Naturally, if she were to keep increasing the sample size, the accuracy of her estimates of the population's true parameters would continue to increase. The only way to ensure absolute accuracy, though, would be to capture every caiman in the lake, an impossible task.

So, Dra. Navideña decides to settle for the n of 10 for the time being and to exploit $\bar{x}$ and s^2 to draw inferences about μ. First, she calculates another sample statistic, the *standard error of the mean*, or $SE_{\bar{x}}$:

$$SE_{\bar{x}} = \frac{s}{\sqrt{n}} = \sqrt{\frac{s^2}{n}} \qquad (5.8)$$

This statistic is a bit different from those presented earlier in that its value decreases as n

increases. The definition is also slightly more complex: $SE_{\bar{x}}$ estimates the standard deviation that would occur not among the individual observations themselves but rather among the independent *means* of a large number of samples each of size *n*, each sample drawn at random from the same statistical population as the sample in question.

According to equation 5.8, the $SE_{\bar{x}}$ of Dra. Navideña's particular sample of 10 caimans is $\sqrt{\frac{0.77}{10}}$ or 0.28. How well does this value, derived from a single sample, actually estimate the standard deviation among a large number of means for samples, all with $n = 10$? Go to box 5.2.

You may be wondering why one would bother to calculate something so esoteric as $SE_{\bar{x}}$ Watch, though, as Dra. Navideña refers to appendix A of this manual and estimates, based only on the statistics calculated from her sample of 10 caimans along with table A.1, the confidence limits for the population mean μ. Making the usual assumptions about random sampling, independence of the different observations, and a normal distribution of values in the statistical population, she now infers that the probability is about 95 percent that the true value for the mean length μ of all adult and

---

Box 5.2. Variation among Means Drawn from the Same Population

Turn to table 5.1 (the lengths of the ninety-nine ferocious caimans), randomly choose 10 individuals (sampling *with replacement*—see note 5, chapter 5), and calculate the sample mean and variance with equations 5.5 and 5.6, respectively. To sample randomly you need some method of generating random numbers between 1 and 99. Many pocket calculators have a random number generator; use the first two digits. Or use a random numbers table such as that found in some statistics texts. Or write the numbers 1 to 99 each on a slip of paper and shake the slips of paper in a bowl, then draw 10 slips blindfolded, returning each one to the hat before drawing again. Whatever your method, *repeat the process 20 times.* That is, draw 20 independent, random samples each of $n = 10$, and for each sample calculate $\bar{x}$ and s^2. If this becomes tedious, convince some friends that they'll have great fun helping you. You won't use your values for s^2 in this particular exercise, but you will use them later.

Now examine the extent of variation among these 20 sample means themselves, by calculating the standard deviation among them (i.e., by applying equations 5.2 and then 5.3 to this set of the 20 *means*). Compare this value to the estimate $(SE_{\bar{x}})$ of 0.28 that Dra. Navideña calculated from her single sample with the same *n*, 10. Are the two values similar?

If you feel especially ambitious, repeat the entire process but with a larger *n*, say $n = 22$ (and more help from your friends). To begin, following Dra. Navideña's example, draw a *single* random sample of 22 values from table 5.1 and calculate $SE_{\bar{x}}$ based on that single sample, using equation 5.8. Is this value lower or higher than the 0.28 she calculated from her sample of 10? Then, again draw 20 independent, random samples, but now each with $n = 22$, from the lengths in table 5.1. For each sample of 22, calculate $\bar{x}$ and s^2 as you did above. Are these means more similar to one another than were those for the samples of $n = 10$? Again, calculate the standard deviation (equations 5.2 and 5.3) among these 20 sample means themselves and compare it to the estimate $(SE_{\bar{x}})$ you'd calculated just based on the first sample of 22. How close are the two values?

Box 5.3. The Risk of Making the Wrong Inference

For each of the twenty samples of $n = 10$ that you drew in the first part of the exercise of box 5.2, calculate $SE_{\bar{x}}$ (equation 5.8) and then, again with the aid of friends if possible, compute the 90 percent, 95 percent, and 99 percent confidence intervals for μ. Did all twenty of the 99 percent intervals include the true value of μ? All twenty of the 95 percent intervals? All twenty of the 90 percent intervals? What does this tell you? In theory, on average two of the twenty 90 percent intervals and one of the twenty 95 percent intervals will miss μ, whereas only one of 100 such random samples will fail to include μ in the 99 percent interval, right? If you've completed the second exercise in box 5.2, now calculate the three confidence intervals for each of the twenty samples with $n = 22$. Has the increase in n led to a narrower confidence interval? But do some confidence intervals—especially at the 90 percent level—still miss the true value of μ?

subadult caimans in the lake lies between 2.0 m and 3.2 m, inclusive.[6] In symbols this is expressed as: $P\{2.0\ m \le \mu \le 3.2\ m\} = 0.95$. Or she might choose to narrow the confidence interval (increase its precision) while increasing the risk that the true parameter is outside that interval (see appendix A), and calculate the 90 percent confidence interval (that is, the interval between the two confidence limits): $P\{2.1\ m \le \mu \le 3.1\ m\} = 0.90$. Note that the risk of missing the true μ indeed increased: this estimate nearly missed what we know to be the real μ, 2.2 m. Alternatively, Dra. Navideña could lower the risk of missing the true μ and accept a fuzzier estimate by calculating the 99 percent confidence interval: $P\{1.7\ m \le \mu \le 3.5\ m\} = 0.99$.

In short, by using a practical definition of confidence intervals, Dra. Navideña can now estimate the mean length of caimans with reasonable confidence (approximately 90 percent, 95 percent, or 99 percent confidence, that is). Of course, she realizes that she always runs the risk of being wrong; her sample could be among the 10 percent, 5 percent, or 1 percent, respectively, of samples in which the confidence interval in fact does not include the true μ. Turn to box 5.3 and see for yourself. With a limited sample, technically speaking, the only way that Dra. Navideña can lower to zero the risk that her confidence interval misses the true μ is to widen that interval to $(0, \infty)$, not a very useful step. The only way she can narrow the confidence limits to the point of specifying the precise value of μ without running any risk of being wrong is to increase n until she has sampled the entire statistical (and biological) population in question. Given an n limited to 10, though, the only way she can narrow the confidence limits is to increase the risk that they do not really include μ, to the absurd point of accepting a P that's near zero if she desires to specify the value of μ precisely—not a very useful step either. So the compromise is to live with a somewhat fuzzy confidence interval that, in practical terms, runs a low but not trivial risk of missing the true value of μ.

The Usefulness of Confidence Intervals

Now you're familiar with one method for relating small samples to big questions: using the sample data to estimate the values of the unknown parameters relevant to the question. Please note that you can expand this approach to calculate standard errors and confidence intervals not only for μ but also for other parameters such as σ^2, σ, population CV, or the slope and intercept of a line.[7] Some leading statisticians who work with field ecology or biodiversity conservation state that confidence limits are a much more valid form of statistical inference than are statistical tests (Yoccoz 1991; Gerard, Smith, and Weerakkody 1998; Johnson 1999). I won't go that far, but I will urge you to calculate confidence intervals whenever they're appropriate to the class of data you've collected and the question you've asked. Also, certain lessons learned from confidence intervals—such as the risk one always runs of being wrong, and the tremendous importance of sample size—extend, and are crucial, to the more complex discussions of statistical inference that follow.

Statistical Inference for Comparative Questions

If you're asking a comparative question, you're wondering whether the design factor has an effect on the values of the response variable. For example, recall the question on selective logging from chapter 4. For any one response variable, such as the number of species of forest birds encountered per standardized evaluation unit within the response unit, you wonder whether a biologically important difference exists between the SL and UL response units (tracts). This question, as always, concerns not just the particular response units in your sample but rather the two statistical populations they represent. In other words, you wonder if $\mu_{SL} \neq \mu_{UL}$, if there's a difference between the means from the two statistical populations consisting of the bird diversities of all possible SL and UL response units in the reserve, respectively. Your statistical concern is whether the two population means might differ from each other. Your conservation concern is that the population means might differ from each other by as much as or more than some quantity δ, where δ is judged to be *biologically significant*.

Most discussions in this chapter, as in chapter 4, will involve questions involving categorical levels of design factors. Likewise, I'll use the concept—and the letter—of δ to signify the true, unknown difference between the means of the different statistical populations of observations (values of the response variable) that pertain to each categorical level, for example between μ_{SL} and μ_{UL}. Nevertheless, similar reasoning applies to the equally common, and equally important, questions that involve continuous levels of design factors. In these cases, your statistical concern is whether variation in the level of the design factor across response units has any effect on variation in the values of the response variable. Your conservation concern is that said effect within the statistical population might be of a biologically significant magnitude e. Technically, the symbol should not be e but rather a Greek letter, shouldn't it? Both δ and e are sometimes called "effect size." For the illustrations that follow I'll stick with categorical levels and δ simply because the logic is clearest that way.

Philosophy

With respect to bird diversity and selective logging, as always you can't ever know μ_{SL} and μ_{UL} for certain. All you know is $\bar{x}_{SL}$ and $\bar{x}_{UL}$, the sample means among the replicate SL and UL response units, respectively. Probably $\bar{x}_{SL}$ and $\bar{x}_{UL}$ differ from each other. But does that difference between sample means necessarily indicate that some real difference δ also exists between μ_{SL} and μ_{UL}? After all, in the exercise of box 5.3 the means that you calculated for the different samples also differed from one another, even though all samples came from precisely the same statistical population of only ninety-nine caimans. Look at figure 5.2, where the real δ is zero but by chance you sampled

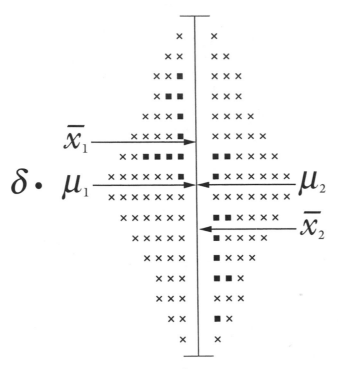

Figure 5.2.
The illusion of difference. Two statistical populations (separated by the vertical line) of observations, with values increasing from bottom to top of the figure. No appreciable difference δ exists between the two population means μ_1 and μ_2 but you don't know that. By chance your sample (blacked-in points) of ten observations from the first population includes an overabundance of its higher values, while your sample from the second population includes a preponderance of its lower values, so that the two sample means differ by quite a lot.

some higher-valued response units in one population and some lower-valued units in the other. If you were to conclude that the two μs were different from one another, you'd be wrong. On the other hand, a real (and biologically important) δ might exist between μ_{SL} and μ_{UL}, but by chance you sampled the lower-valued response units from the level with the higher μ and the higher-valued units from the level with the lower μ, as in figure 5.3. You've ended up with a minuscule difference between $\bar{x}_{SL}$ and $\bar{x}_{UL}$ and, unable to discern the true δ, cannot infer that the design factor has any effect on the values for the response variable.

Do the chances of making one or the other mistake depend on sample size? Of course. If you naïvely used design 7 to address the question on selective logging, you'd almost certainly find a difference between the two values. Again, the numbers of response units are ridiculous: $n_{SL} = 1$ and $n_{UL} = 1$. Based on these sample sizes, would you conclude confidently that a worrisome δ exists between μ_{SL} and μ_{UL} throughout the reserve's perimeter, and base management decisions on that conclusion? I hope not. On the other hand, a vast δ could really exist between μ_{SL} and μ_{UL}, but with sample sizes of 1 and 1 you'd be unable to blame the observed difference between these two values on the design factor because you can't honestly say that intrinsic variation wasn't responsible (see chapter 4).

Let's say that instead you've more wisely employed design 12, and that with $n_{SL} = 6$ and $n_{UL} = 6$ you find that $\bar{x}_{SL} > \bar{x}_{UL}$. Indeed, you find that in every one of the six blocks, the number of forest

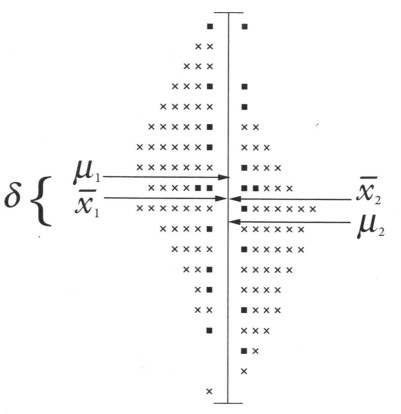

Figure 5.3.
The illusion of sameness. As in figure 5.2, but this time a substantial difference δ really does exist between the two population means μ_1 and μ_2. You don't know that, though. By chance, you've sampled a similar distribution of values from each population, so that the sample means do not differ appreciably.

bird species in the SL plot exceeds the number in the UL plot. Now can you conclude with absolute confidence that $\mu_{SL} > \mu_{UL}$ and proceed with management guidelines? That is, can you infer that if you were able to continue sampling forever throughout all the theoretically possible UL and SL tracts in the reserve, this inequality would continue to hold? Not necessarily. After all, even with six replicates per level there's still a chance that your sampling resembles that shown in figure 5.2. The greater the intrinsic variation within each of the two statistical populations, the more difficulty you'll have drawing inferences, based on your samples alone, about the true nature of μ_{SL} and μ_{UL}. The larger those samples, the less difficulty you'll have, though.

Inference for Nondirectional Questions

Really, in many cases, answering comparative questions through statistical inference boils down to an attempt to distinguish effects of the design factor on the nature of your samples from the effects of intrinsic variation within the statistical population(s) of values. Note how important this attempt is to your conservation concern, conservation guidelines, or basic field study. Wouldn't it be nice to have a more objective means of distinguishing those two sources of variability within a given response variable? *Statistical tests* are just that.

To begin a statistical test, before collecting the data, you must formalize your dilemma as two *statistical hypotheses* regarding the response variable in question. These are, of course, the statistical null hypothesis H_0 and the statistical alternative hypothesis H_A (see figure 2.1). The two hypotheses follow the general form of:

H_0: There *is no* effect of the design factor on the values of the observations among the statistical population(s) in question.

H_A: There *is* an effect of the design factor on the values of the observations among the statistical population(s) in question.

In classical statistical inference as it's commonly used, all that you'll do after collecting the data is evaluate H_0 and decide whether to reject it. That is, you'll assess the possibility that the apparent pattern among your observed data, or patterns even more extreme, could have been generated by the vagaries of sampling from statistical population(s) where the H_0 actually held. As usual, your evaluation can never be 100 percent certain. That is, if based on your sample data you decide to reject H_0 and infer that the design factor indeed plays a role, you always run the risk of being wrong without knowing it, which is called a *Type I statistical error*. For example, a researcher who collects the data of figure 5.2 and decides to reject H_0 is unknowingly making a Type I error. On the other hand, you run the risk of being wrong in the opposite direction, or making a *Type II statistical error*, whenever you decide *not* to reject H_0. The researcher who collects the data of figure 5.3 and decides not to reject H_0 has just made a Type II error. This clear dichotomy between Type I and Type II errors is the crux of practical statistical inference and, more important, the crux of conservation decisions based on inquiries, as you'll see.

To summarize: *A Type I error is deciding to reject the null hypothesis when you really shouldn't have rejected it, and a Type II error is failing to reject the null hypothesis when you really should have rejected it.* Of course, you'll never know whether you're committing such an error. All you know is that whatever your decision about the null hypothesis, there's always a risk of being wrong. That's life.

Traditionally, researchers in our field and others have been much more afraid of committing Type I than Type II errors. The concept of Type II errors gets a brief mention, at best, in many statistics tests and most statistics courses. Why the disparity? Intrinsic variation and the vagaries of random sampling almost always create illusory patterns in the data (see figure 5.2). So before ascribing those patterns to the design factor, most people want to be very confident—recognizing that they can never be 100 percent confident—that they're reducing the risk of a Type I error to an acceptable, low level. If they are following the usual practice, *before* collecting data, researchers must decide objectively on the highest level of α they're willing to risk, where α is now defined as the probability of committing a Type I error.

Here's how this method works. Let's say that following your study, the data analysis shows that, if you were to decide to reject the null hypothesis, the risk of committing a Type I error would exceed this predetermined *rejection level* or $\alpha_{rejection}$ (hereafter abbreviated as α_{rej}). In such a case you must not reject the null hypothesis, no matter how much you're tempted to do so; you must play by the rules that you yourself set up before the study. If the analysis shows that the risk is equal to or less than α_{rej}, you may decide to reject the null hypothesis. As in the selection of confidence intervals, the default rejection level tends to be $\alpha_{rej} = 5\%$, or 0.05. There's nothing magical about 0.05, though, other than that most of us have five digits on a hand, and thus 5 (actually, 10—both hands) is the basis of our number system. If you're willing to run a greater risk of committing a Type I error, for reasons detailed below, you might have selected 0.10 or an even higher α_{rej}. Conversely, if the social, health, conservation, or economic cost of committing a Type I error is especially severe, you might have chosen a conservative α_{rej} of 0.01 or less.

Once you've made the decision about the rejection level, you forget all about that for the time being and proceed with data collection. Afterward, you enter the sample data into a mathematical formula particular to the statistical test chosen. The equation produces a value, the *test statistic,* particular to your data set. You next turn to a table for that class of statistical test. Most (not all) statistical tables vaguely resemble table A.1, with columns corresponding to certain values for α, rows corresponding to *degrees of freedom* or *df,* and a set of *tabulated values* for the test statistic. Don't worry now about the meaning of degrees of freedom. Pragmatically speaking, it's a number that's determined by sample size and by the particular test used.

You now use the table to find the α that's actually associated with the data you've just collected, or $\alpha_{observed}$ (α_{obs} from now on). Often α_{obs} is called the *P*-value, but its meaning and numerical value are very different from those of the *P* of confidence intervals. You enter the table at the row that corresponds to the number of degrees of freedom, seek the location of your test statistic among the tabulated values, then record the α associated with the tabulated value that matches your test statistic. This value, α_{obs}, is the probability that an observed value for the test statistic equal to or greater than the particular value you obtained could have arisen just through the vagaries of sampling from statistical populations that in truth conformed to H_0. If you decide to reject the null hypothesis, then α_{obs} is the precise risk you're running—given the usual assumptions—of having just committed a Type I error. If you're using a statistical table, only rarely does the value of your test statistic precisely match any tabulated value in the appropriate row, so you can only bracket α_{obs}, for example, $0.002 < \alpha_{obs} < 0.005$ (alternatively, $0.002 < P < 0.005$). Statistical packages on computers, though, usually provide the precise value for α_{obs} (P).

You should view α_{obs} from two perspectives. First and most important, as just described, α_{obs} provides the exact probability that the test statistic or one more extreme could have been generated by sample populations where H_0 were true, that is, the probability of committing a Type I error should you reject the null hypothesis. Thus, α_{obs} is a critically important number whose precise value should be reported along with the value of the test statistic itself, the sample statistics (for example, $\bar{x}$ and s^2), the n for each of the k levels of the design factor, and if appropriate the confidence intervals for the parameters of interest. Second, though, if you're following the "rejection level" approach, you compare α_{obs} with the value you'd previously chosen for α_{rej}. Now's the time to be absolutely honest. Again, an $\alpha_{obs} \leq \alpha_{rej}$ permits you to reject H_0, whereas if $\alpha_{obs} > \alpha_{rej}$, you cannot. In the first case, in particular if you'd chosen the "default" α_{rej} of 0.05, you might say that the results are *statistically significant.* You should recognize, though, that this label is simply a convention for the default acceptable risk (5 percent) of committing a Type I error, a convention arrived at because of the "five-finger syndrome" mentioned above. Ponder how different the conclusions of some ecology or conservation studies might be if centuries ago most humans had had three, or seven, fingers per hand.

Different Inference Rules for Directional Questions?

Most comparative questions simply wonder whether the levels of the design factor have an effect of any sort—for example, either positive or negative—on the values of the response variable. They're *nondirectional* questions. All versions of the selected logging question in chapter 4, and all of the real-life questions in chapter 3, were nondirectional. To analyze results from a study based on a nondirectional question, when following the process described above, we apply what are called "two-tailed" statistical tests and record what are called "two-tailed values" for α_{obs}. Usually, this doesn't require a conscious effort: the text already provides the equation for a two-tailed test, and the table already gives the two-tailed values for α. Likewise, I assumed above that two-tailed processes were being followed.

Occasionally, a researcher has a strong reason for suspecting, or fearing, that the design factor will have an effect in one direction only. He or she might wish to elaborate a *directional* question, such as "In the periphery of the reserve, do SL parcels have fewer species and individuals of understory frogs than do UL tracts?" In our fields and others, such as the behavioral and social sciences, researchers often follow up directional questions with a slightly different process for inference, consisting of "one-tailed" statistical tests or "one-tailed" values for α_{obs}. Why would one ever do this? The pragmatic answer: If you're pursuing the Holy Grail of statistical significance, for a given data set it's twice as easy to achieve statistical significance with one-tailed statistical tests or one-tailed values for α_{obs} as it is with two-tailed tests and two-tailed values.

The use of one-tailed testing procedures is highly controversial. Many statistics texts and some statisticians continue to propose that one-tailed testing procedures are valid in themselves, if used properly. Other statisticians assert that they are valid only under very special circumstances that almost never occur in field inquiries, whether on basic or applied questions (S. H. Hurlbert, personal communication). Whatever their technical validity, in practice one-tailed procedures are seldom used properly, partly as the result of confusion created by some statistics texts and tables (see below). There's no urgent reason ever to use one-tailed procedures, even if you were to do so with the utmost care and honesty. Taking all these considerations into account, it seems best to obey the following: *Don't ever use one-tailed tests or one-tailed values for* α_{obs} *(i.e., one-tailed P-values)*. Be especially careful when you refer to the text by Siegel and Castellan (1988, 1995): some statistical tables there present values for two-tailed procedures, some for either one- or two-tailed procedures, and still others for one-tailed approaches only. If you encounter a statistical table that provides only one-tailed values for α_{obs}, whether in Siegel and Castellan or another text, all you need to do is double all of them (for example, a column headed by "$\alpha = 0.05$" would become "$\alpha = 0.10$"). You would then have a table for two-tailed tests, as described in the previous section.

All these discussions of Type I errors, α, one- and two-tailed α-values, and rejecting or failing to reject null hypotheses may seem terribly trivial, esoteric, and confusing. Nevertheless, you must recognize that the procedures for statistical inference that we've discussed have played an overwhelmingly dominant role in both basic ecology and biodiversity conservation for many years, and that they show no signs of letting up. "Statistical significance," that magic value of 0.05, dictates a lot more conclusions and decisions than it has any right to do (Yoccoz 1991; Crome 1997; Johnson 1999). Still, statistical tests can be viewed and used correctly; and when they are, they provide a valuable philosophical, logical, and practical tool.

Recipes for Statistical Tests

If you want to apply a statistical test to some real data, go elsewhere (see note 1 of this chapter), and be careful. A tremendous variety exists of parametric statistical tests, corresponding to the vast array of study designs available. Some tests are quite simple, for example, the well-known two-sample t test, possibly useful for design 11 of chapter 4, or the paired-sample t test, possibly useful for design 12. Others are extraordinarily complex. The numbers of complex tests grow exponentially as computer packages become increasingly sophisticated. All parametric tests, though, make certain assumptions regarding the nature of the statistical population, the nature of the response variable, and the sampling procedure. I'll return to this point later in the chapter.

Beyond α: Making Statistical Inference into a Powerful Conservation Tool

As a conservation scientist or manager you should really fear making a Type II error at least as much as making a Type I error (figure 5.4). Often the conservation concern involves a potential threat to the integrity of your protected area or the possibility that some choices of guidelines will affect the reserve less benignly than others. If you undertake an inquiry, perform an observational or experimental study, and don't reject H_0, you cannot legitimately infer that the design factor—in this case, the threat or the choice of guideline—is anything to worry about. What if, though, by making this decision, you'd unknowingly committed a Type II error? If you were rigorously following the scheme of figure 2.4 by means of the "α-heavy" approach to statistical inference that's traditional in our fields, you'd go along blithely ignoring the potential threat to your reserve's integrity or the

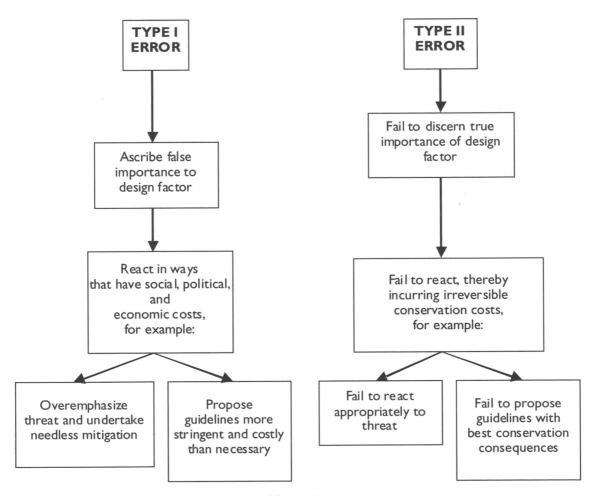

Figure 5.4.
The costs to conservation of committing either a Type I or a Type II statistical error. Which side, the right or the left, concerns you more? Modified from Noss (1991).

potential, biologically significant difference in the outcomes of alternative guidelines. The risk to biodiversity conservation of committing a Type I statistical error needs to be balanced against the risk of committing a Type II error. Your problem is, though, that you can't minimize both risks at the same time except by increasing your sample size to infinity. What's a reasonable compromise?

Type II Errors and Statistical Power

We've defined α as the risk of committing a Type I error if you decide to reject H_0. Let's define β as the risk of committing a Type II error (see figure 5.3) if you decide not to reject H_0. Like α, in theory β can vary from 0.0 to 1.0. At worst, the design factor could actually be having a very strong effect within the statistical populations in question, but your study is so weak that you're unable to reject H_0 no matter what (that is, $\beta = 1.0$). What's an acceptable value for β that you should seek to achieve in your study design? Is it 0.01, 0.05, 0.10, 0.20, or 0.40? The choice for β as for α depends on you. Unlike the case of α, though, no convention exists, although the recent literature in conservation and management suggests possibilities.[8]

For a given set of n data, β and α are inversely related. By insisting on a very low α do you increase the risk of committing a Type II error when you decide not to reject H_0? If you relax α so as to lower β, do you increase the risk of committing a Type I error when you decide to reject H_0? Is your chance of distinguishing any effects of the design factor from the "background noise" of intrinsic variation greater when the true (and unknown) δ (or e) is large than when it's small? Would the two risks, α and β, differ between a study with n replicate response units per level and a study with exactly the same values for $\bar{x}$ and s^2 as the first but with $3n$ replicate response units per level? The answer to all is yes. Please ponder these questions a bit, referring again to figures 5.2 and 5.3, before reading on.

Let's define a mathematically simple but conceptually crucial term: the *power of a statistical test* $= 1 - \beta$ for a particular data set. For inquiries having categorical levels of the design factor, the power of a test is thus the probability that, using this test, you would reject H_0 if a difference of size δ or greater really did exist among the statistical populations of observations for those levels. In other words, the power of the test is the probability that you'll reject H_0 when in fact you should do so. Four factors influence the power of a test:

1. For a given sample of data, power increases as you relax α_{rej} (i.e., allow the risk of committing a Type I error to increase).

2. For a given sample of data, power increases with increasing magnitude of δ (or other effect of the design factor) among the statistical population(s) in question.

3. For a given α_{rej} and δ, power increases with increasing n.[9]

4. In cases where the data could be analyzed with more than one kind of statistical test, power may vary with the particular test chosen.

Too Little Power or Too Much?

Let's see where these four factors lead in practice. If you were a mercenary ecologist employed to do environmental impact assessments for a gold-mining company pouring mercury and cyanide into the local watershed, you'd recognize that a substantial true δ almost certainly existed in water quality between affected and nonaffected streams. How might you *decrease* statistical power, thereby increasing the chance that you'll fail to reject H_0 and gratify your employers, who can then say, "See? There's no statistically significant effect of our activities!"? You could decrease α_{rej} and/or keep n very small and/or apply the weakest possible kind of statistical test. Does this scenario ever really occur? Undoubtedly, and you should be aware of that possibility. Now, do you yourself want to decrease statistical power in this fashion? I certainly hope not. Rather, you should seek reasonably high statistical power. But there may be a limit to that, also. You need a large enough n to provide sufficient power to detect a biologically significant δ, but you don't need so much power that you'll end up detecting any δ, no matter how microscopic it may be.

For example, Dr. Félix Cumpleaños worries that agricultural runoff into unprotected Laguna Encontrada might be seriously stunting the growth of caimans there, in comparison with caimans in the pristine Laguna Olvidada already investigated by Dra. Navideña. Note that he's just comparing two sites; his design is not meant to evaluate the effect of agricultural runoff per se. Dr. Cumpleaños undertakes an exhaustive sampling program and measures 72 of the underlying 99 caimans in Laguna Olvidada (table 5.1) plus 427 of those in Laguna Encontrada. With this tremendous sample size the α_{obs} is even lower than the α_{rej} of 0.01 that he had chosen beforehand. Dr. Cumpleaños announces truthfully that the caimans of Laguna Encontrada are significantly smaller, *statistically speaking*, than those of Laguna Olvidada. Actually, though, in his data the observed difference between $\bar{x}_{Laguna\ Encontrada}$ and $\bar{x}_{Laguna\ Olvidada}$ is 1.2 mm, and the true δ between $\mu_{Laguna\ Encontrada}$ and $\mu_{Laguna\ Olvidada}$, unknowable to him, is only 1.3 mm. His results may be statistically significant, but are they biologically significant? They seem biologically trivial to me. On the other hand, the unscrupulous ecologist working for the gold-mining concern may succeed in failing to discover statistically significant differences in water quality but in private admit that the "insignificant" difference is actually of tremendous biological significance.

In short, what's a reasonable compromise between a sample size too small to be able to detect a biologically important δ or e, and a sample size so large (and costly) that it'll nearly always assign "statistical significance" to a data set from statistical populations with a ridiculously small, biologically insignificant δ? See appendix B, which exploits the logical relationship between four variables— α, β, n, and δ—to come up with crude estimates of the compromise sample size you'll need. If at all possible, you should apply the technique of appendix B or a better one to a set of preliminary data *before* performing your study, if there's any latitude at all in the number of replicate response units available to you. That is, you should decide on the sample size per treatment level that, with a reasonable α, will likely provide you with adequate power (say, 80 percent or 90 percent) to reject H_0 if there really is a δ large enough to be biologically significant. How, then, do you choose that biologically significant δ that you wish to be able to detect at least $(1 - \beta)$ percent of the time? No statistics book will tell you, nor can I. You must decide on the magnitude of change in the value of the response variable that signals something interesting, or worrisome.

Perhaps you'll go through the exercise of appendix B and discover that the required n is so high that you cannot afford the time or resources to complete an adequate study. Or you might be limited to a fixed n, and, by using appendix B but solving for β rather than n, you find that the available n gives you very little power to reject H_0 if the δ you feel to be biologically significant truly exists.[10] More likely, though, you'll come out with a minimum sample size that's reasonable to achieve, and that will allow you to draw inferences with a reasonably low risk that you're committing either a Type I or a Type II statistical error.

In short, as appendix B demonstrates, you can use statistical inference in one sense (estimating minimum sample size) to guide you to a study design that's adequate for statistical inference in another sense (calculating confidence intervals, or deciding whether to reject H_0 and then proceeding with reasonable confidence toward reflection and application). More and more conservation professionals, as well as basic scientists, recognize the crucial role of considering statistical power when making decisions.[11] More and more recognize that the risk of Type II error may be much more serious to long-term conservation concerns than the risk of Type I error (see figure 5.4) and decide reluctantly to relax α_{rej} in order to increase power in cases where n has to be limited. Whether or not you apply these procedures to your own field studies or conservation decisions, at least you should recognize the logical interplay between the two risks α and β sample size and the true effect that the design factor is having on the response variable in question. Failure to reject H_0 is not the same as showing that the design factor has no effect. Far from it.

Nonparametric Statistical Inference: Often the Better (and Simpler) Choice

Most features of descriptive statistics or statistical inference discussed so far assume that your response variable involves a statistical population of values that are continuously distributed (i.e., an observation could take any value, including fractional values). Furthermore, all procedures discussed so far rely on the concept of the parameters μ and σ^2 and require that estimates $\bar{x}$ and s^2 can be calculated with some degree of meaning. Many of the more complex parametric statistical tests further assume that the population variances among different levels of the design factor are equal. Not all response variables meet all these criteria. Even the legendary caimans of table 5.1 were measured only to the nearest 0.1 m, so that the observed values, at least, do not technically present a continuous distribution.

In many cases, the response variable deviates much further from the assumptions of popular parametric tests. For example, let's say that Dr. Cumpleaños is concerned not only that agricultural runoff may have influenced the size of caimans but also that pesticides in the runoff may have affected endocrine processes and thereby the sex ratio of subadult and adult caimans. He wishes to compare the sex ratio of caimans in Laguna Encontrada with that of caimans in protected Laguna Olvidada. Again, his question involves the two lakes only and is not intended to address the question "Do sex ratios of caimans differ between lakes that are exposed to agricultural runoff and lakes that are not?" Once again Dr. Cumpleaños samples a large number of caimans but this time only to record the sex of each. The response variable has two possible measures: male and female. For each lake, can he calculate $\bar{x}$ and s^2 among these measures? No. Does the response variable display a continuous distribution of values? No. These data have little to do with parametric statistics. Instead of a theoretically infinite number of categories for the response variable, just two exist, and these can't realistically be ranked in value.

Other cases where parametric statistical inference is inappropriate may be less obvious. Dr. Cumpleaños has still another worry: that agricultural runoff into Laguna Encontrada has affected the proportions of caimans among the different age classes, as compared to the caimans of Laguna Olvidada. His question is "Does the distribution of individuals among the different age classes differ between the two lakes?" His response variable can take on any one of four age classes: recent hatchling, yearling, subadult, and adult. He returns to the lakes to sample caimans, not just the two oldest age classes but also the two youngest ones. He assigns each individual captured a code for its age class, respectively 1, 2, 3, or 4. That is, the response variable involves only these categories rather than a potentially infinite number.

In contrast to the question on sex ratio, this time the categories can reasonably be *ranked* with respect to one another. A caiman with code 4 is unequivocally older than one with code 3. Are these data appropriate for parametric statistical inference? Not at all. Obviously, the caimans are not aged precisely to the day, week, or month. They are simply assigned to arbitrary categories. Almost certainly, many ties will occur among the observations—that is, if Dr. Cumpleaños obtains a decent sample, there will be many individual caimans each given rank 4, many given rank 3, etc. It makes little sense to calculate the mean, let alone the variance, among these ranks. There's no reason to expect ranks in the relevant statistical population to resemble a normal distribution.

In short, many perfectly valid studies involve perfectly valid response variables that are quite inappropriate for parametric statistical analyses but are entirely appropriate for *nonparametric statistical inference*. Consider the three different kinds of response variables we've discussed. Measurements of the length of caimans exemplify, or nearly exemplify, *interval data*. The scale that's used (to 0.1 m) measures quite precisely the interval between the lengths of any two caimans, and that interval (as well as the measurement of any one length) could have a unique value. If a set of interval data complies with the remaining assumptions required by parametric statistical inference in general and the test chosen in particular, then you can proceed with a parametric test if you so wish.

The age-class data compiled by Dr. Cumpleaños exemplify *ordinal data*. The categories for the response variable can be ordered, or ranked from small to large or low to high, with respect to one another. Multiple observations within a given category, though, cannot be further ranked. For instance, there's no way of giving different values to two different individuals both having rank 2. The categories themselves are sometimes, although not always, arbitrary. The breadth of one category is not necessarily the same as the breadth of another. For example, category 1, "recent hatchlings," may include a much narrower range of true ages and lengths than does category 4, "adults." Depending on the nature of the categories, a parametric statistical test performed on ordinal data is somewhat, quite, or ridiculously inappropriate.

Finally, "sex of caiman" exemplifies a response variable whose categories cannot be objectively ordered or ranked with respect to one another, or *nominal (classificatory) data*. Each caiman (the response unit) can be assigned unequivocally to one or the other category, male or female, such that each category accumulates a count or *frequency* of caimans assigned to it. The response variables for many key inquiries on biodiversity themes involve nominal data. For example, you might ask whether selective logging is associated with a change in frequency of the different species of small mammals. The response variable that you "measure" on each individual captured is "species of small mammal." You cannot logically rank in order, or calculate the mean and variance of, "species." Such data absolutely require a nonparametric procedure.

Most standard statistics texts pay lip service to nonparametric statistical inference, and Zar (1999) does a truly excellent job of presenting both parametric and nonparametric approaches, but you should really consult more complete resources if possible.[12] Many response variables for the questions most important to management concerns involve nominal or ordinal data and require nonparametric rather than parametric approaches. Many sets of interval data that technically fit the criteria for parametric inference can also be analyzed with nonparametric tests. Although parametric tests for complex study designs have (as yet) no analogue in the nonparametric world, for many of

the simpler designs there exist nonparametric equivalents to the parametric tests that would traditionally be applied. Among numerous kinds of nonparametric procedures for different study designs,
some of my favorites, lucidly described in Siegel and Castellan (1988, 1995) and elsewhere, include:

- the chi-square one-sample test, for example to compare the observed frequencies of male and
  female caimans to an expected frequency of 1:1;[13]

- the chi-square two-sample test (test for heterogeneity or association), useful for the comparison
  by Dr. Cumpleaños of the sex ratios in the two lakes (this can also be extended to more than two
  samples and more than two levels of the design factor);

- the Kolmogorov-Smirnov two-sample test, useful for the comparison of age-class distributions
  between the two caiman populations;

- the Wilcoxon signed-rank test for paired samples, useful for design 10, 12, or 13 of chapter 4 if
  the ns were larger (also the Friedmann test for designs with more than two categorical levels);

- the Wilcoxon-Mann-Whitney "U" test for two independent samples, useful for design 11 of chapter 4 if the ns were larger (also the Kruskal-Wallis test for designs with more than two levels);

- the Siegel-Tukey and Moses tests to compare the amount of variability between two samples;

- the Spearman rank correlation test, useful for examining the relationship between two response
  variables each measured across a number of response units that themselves constitute the design
  factor;[14] no examples have been presented, but one case might be the number of parasite species
  in the stomachs of caimans and the number of biologists consumed per year, the design factor
  being simply "caiman."

If you have the choice of applying either a parametric test or its nonparametric equivalent to your
particular design, I'd usually suggest the nonparametric approach. Nonparametric inference has
many advantages. Besides the considerations mentioned above, one practical consideration is that
you can perform almost any nonparametric test easily with a simple hand calculator. Recognize,
though, that nonparametric approaches aren't always ideal, as follows:

- Since nonparametric tests don't involve the estimation of the population mean and other parameters, an appropriate analogue to δ or e (effect size) isn't always obvious, although see Cohen (1988).

- Again, some designs can be analyzed only with parametric approaches. One example is design 16
  in chapter 4. If your question calls for a complex design and you can reasonably justify the assumptions for a parametric test, use the parametric test. In this case, if you tried to cram the data into
  a nonparametric test appropriate for a simpler design, you'd lose information and precision.

- In some cases, where data are appropriate for either one, statistical power decreases when a nonparametric test is used in place of the parametric equivalent.

- Parametric tests are widely used and well known, such that many (not all) investigators apply and
  interpret them correctly, whereas nonparametric tests are more often misused and misinterpreted
  than not. I have seen the simple chi-square test used incorrectly far more frequently than used correctly.

- Some investigators believe that they can apply nonparametric tests when they've failed to meet the criteria of random sampling or independent data, when the data were measured poorly, or in other cases of poor study design. They're dead wrong.

Other Statistical Approaches

The "computer age" has enabled statisticians to develop and disseminate many sophisticated statistical techniques that now are often used, or encountered, by managers and other conservation professionals. Most techniques fall into four broad classes.

Multivariate Statistics

This class includes many statistical techniques that involve several or many response variables at a time, perhaps correlated with one another, and in some cases several design factors as well. These include Multivariate Analysis of Variance (MANOVA), Principal Components Analysis (PCA), Detrended Correspondence Analysis (DCA), Discriminant Function Analysis (DFA), Canonical Correspondence Analysis (CCA), and many others. Note the prevalence of acronyms, and recall the dedication to this book. Multivariate statistical techniques can be powerful tools for conservation questions—if used properly. They're often used improperly.

For example, some techniques, such as MANOVA, are indeed intended for statistical inference. Nevertheless, they involve so many assumptions (rarely tested) about the nature of the data and are so sensitive to slight aberrations in the data or to outlying values that the quantitative results they produce should be viewed with some skepticism (James and McCullough 1990). One useful technique that analyzes nominal data in three or more dimensions, log-linear analysis, seems to allow for statistical inference but really is best used for initial explorations of the nature of the data set.[15]

Other multivariate techniques, such as PCA, DCA, and CCA, are strictly descriptive, contrary to the hopes (and interpretations) of many people who use them (James and McCullough 1990; McCune 1997). They can be extraordinarily useful as exploratory techniques. Most are designed to illuminate the underlying patterns in a very complex data set such as counts per species in samples of vegetation or in assemblages of small mammals. They really just produce fancy descriptive statistics, though. As such, they're very sensitive to the precise nature of each response unit that's included. Adding or subtracting a single unusual response unit can markedly change the entire outcome. Most multivariate techniques lose their credibility if the data set includes many zeroes. Also, different techniques—or different choices for the particular mathematical algorithms available within any one technique—often produce very different results when applied to the same data set. Many users simply plug their data into the most convenient computer program without further questions and without recognizing these concerns. Excellent, if overly enthusiastic, introductory texts exist on multivariate techniques, but you should read the superb review by James and McCullough before using, or interpreting, any multivariate technique that's labeled with an acronym.[16]

Resampling Statistics

Some fairly new approaches to statistical inference employ computer-intensive methods to obtain estimates of the parameters of the statistical population, or evaluations of the statistical null hypothesis, that the sample itself generates. The computer reshuffles, or resamples from, the original observations a very large number of times, according to the philosophy and technique of the particular approach. These approaches include "bootstrap," "jackknife," "Monte Carlo," and "randomization" procedures. Such procedures can be highly useful to some conservation questions as well as to basic inquiries. I won't describe resampling statistics further, leaving that up to several excellent references.[17] As always, be extremely cautious when using computer-intensive methods. For example, some procedures compare patterns among the observed data with those produced by a "null model" that's generated by the resampling process. Coming up with a reasonable null model is the most difficult part, and critics point out that so-called null models often include hidden patterns, restrictions, and biological assumptions.

Meta-Analysis

What if you yourself lack adequate time to perform a well-designed study on the question critical to your conservation concern? You might use the technique of meta-analysis and synthesize the results of all similar studies that you can find in the literature, no matter who performed them or where they took place. Through this technique you can estimate the typical δ, or more generally the effect size e, associated with the design factor (for example, selective logging) throughout the geographic range covered by the set of published studies you've used.[18] Exercising great caution, you might assume that this universal effect size applies in your landscape as well and use the result of your analysis to propose new conservation guidelines locally.

In theory, meta-analysis could be extremely useful to conservation professionals and to other researchers as well. Unfortunately, though, by definition meta-analysis depends on the work of a lot of other people, and there's a very real tendency among basic and applied ecologists to publish results only where H_0 was rejected and the results conformed with their expectations or advocacy positions (see Johnson 1999). Therefore, no matter how diligently and objectively you yourself search for published studies, meta-analysis will almost certainly overestimate δ or e and lead you to overemphasize the influence of the design factor, whatever that may be. Perhaps meta-analysis is best left for the biomedical and social sciences where it originated and where it's just as easy (and acceptable) to publish results that contradict the paradigm of the moment as to publish those that support it.

Bayesian Statistics

The Bayesian approach to statistics was proposed in the eighteenth century but is just now beginning to catch on among field investigators. For most of us, Bayesian methods represent a novel approach to statistical inference, very different from the methods of the classical approach (termed "frequentist statistics" by Bayesian statisticians) discussed here and still employed by most basic statistics texts, statistics courses, and published studies on basic or applied ecology. The Bayesian approach holds great promise for questions in conservation and management (Wade 2000). In essence, one begins a process of Bayesian inference not by proposing null and alternative hypotheses but instead by proposing a theoretical value for, or belief about, the effect size (e or δ) or parameter in question. Following the study, data are then used to modify that belief or theoretical value and to propose a new value, or a new distribution of the probabilities associated with possible values, with which future studies are evaluated in turn. Results from a single study, even with the n quite small, can be used to set probabilities on the outcomes of future studies, or events, taking place under the same conditions. Bayesian statistics bypass any confusion between statistical and biological significance; indeed, they address biological significance—effect sizes, parameter values—directly. Most ecologists, conservation biologists, and their professors have been raised on an exclusive diet of classical statistics. Undoubtedly it will take some time for the Bayesian cuisine to catch on with the majority. Nevertheless, if complex equations don't bother you, at least acquaint yourself with the basics of Bayesian methods.[19]

What to Do?

What's the best way to apply statistical inference to field ecology and to conservation concerns, inquiries, and decisions? As usual there's no one answer, but a few suggestions follow.

1. Consider whether you even need a statistical test. Recall the purpose of statistical inference (and see point 8 below): to extrapolate from a limited number of data to the larger scope of all possible samples that could be drawn under the same conditions. Is this always what you need? Many times statistical inference is unnecessary or even inappropriate to your particular goal. In some cases practical considerations or ethical concerns must limit sampling to such an extent that inferential statistics make no sense. In a few cases the sample alone holds your interest; for example, you don't always need statistical support to demonstrate that a design factor has strong, biologically significant effects. Recognize that good study design (chapter 4), which is really just common sense, is often more important than the fine details of statistical inference. Again, some authors (e.g., Yoccoz 1991; Johnson 1999) point out that statistical testing is used properly with such low frequency that perhaps it should be thrown out entirely, except for estimating population parameters. Before gleefully ripping this chapter out of the book and stomping on it, though, pay heed to this quote from Crome (1997): "Don't misuse statistics. Just because statistics are usually misused, misinterpreted, or misunderstood doesn't mean that they are bad methods, and it doesn't mean you should do the same."

2. Have at hand not only a standard statistics text (see note 1 for this chapter) but also Siegel and Castellan (1988, 1995) or another text on nonparametric techniques, and use each book with caution.

3. When appropriate, though, calculate confidence limits on the parameters of interest for the different levels or samples. These alone will give you a good idea of whether the design factor is likely to have biologically significant effects.

4. Use the simplest test that fits your design. In general, the more complex the test, the more assumptions it makes about your sampling and the nature of the data in the statistical populations concerned—thus, the less you should trust the quantitative outcome.

5. Likewise, for reasons mentioned above, approach cautiously any of the "sophisticated statistics" described in the previous section. Often you'll be encouraged, or tempted, to apply the most sophisticated technique possible. Resist.

6. When in doubt, consult a professional statistician. Recognize, though, that most professional statisticians deal with well-controlled designs for experimental studies in the agricultural or biomedical sciences. They are often unaware of the complexities in the design of field studies, the crude nature of many field data, the lack of control over possible confounding factors, the fre-

quently arbitrary selection of evaluation units, the fact that response units often differ from evaluation units in ways not immediately apparent, and the many data sets that include numerous zeroes. Also recognize that there's great divergence of opinion among the leading statisticians. Ask ten statisticians the same question, and you'll get back ten quite different answers. Seek statisticians' advice without hesitation and be grateful for it, but evaluate their suggestions carefully before applying them. Reread chapter 4 before your appointment, and perhaps afterward too.

7. Don't dismiss the philosophy. Even if you're feeling overwhelmed by the onslaught of Greek symbols and equations and vocabulary in this chapter, even if you never intend to let a desktop computer sully your office, recognize that a basic understanding of the fundamental philosophy of design (as presented in chapter 4) and that of Type I and Type II statistical errors (as presented here) is indispensable to making careful decisions on conservation and management themes, or to drawing careful conclusions from your field study. Recall that whatever you say or do, you run the risk of being wrong. In particular, you always run the risk of making a Type I error if you decide to say that a pattern really exists (rejecting H_0), and you always run the risk of making a Type II error if you decide not to say that a pattern exists (not rejecting H_0).

8. Reflect once more on the rationale for statistical inference: the relationship between a sample of certain values and a (much) larger set of data that could potentially be collected under exactly the same conditions. How reasonable is that innocuous little phrase "collected under exactly the same conditions" in practice? Do you really think that you could ever encounter a second instance with exactly the same conditions? It's time to leave the Greek symbols (at least most of them) and go back to natural history.

CHAPTER 6

Points of View: Taking Natural History into Account

Acts in . . . the "ecological theatre" are played out on various scales of space and time. To understand the drama, we must view it on the appropriate scale.

—John A. Wiens (1989)

You've just been cautioned to take special pains with study design, to reflect carefully on just how far you can extend an inquiry's conclusions, and to ponder just how realistic is the "fundamental rule of statistical inference." Why? Because in the landscape where you work, nothing remains the same from one place to another or one time to another. That is, if you rely blindly on statistical inference from samples taken in one place and at one time for applications to other places and other times, there's one basic problem: the statistical population from which you drew the sample no longer exists.

Furthermore, every living organism has its own point of view on space and time. That point of view relates to all other aspects of the organism's natural history. For example, your spatial point of view as a human being normally resides at about 1.6 m above the ground, and encompasses about a hectare in area. That is, you take special note of the circumstances within a radius of about 50 m (or less, in a dense forest). Thus, as a field ecologist or conservation professional, you also tend to use that scale when you think of the "site," "stand," "patch," or "ecological community"—about 1 hectare. Can you rely on this human point of view on natural heterogeneity to guide you toward the best designs for inquiries and the guidelines that best ensure long-term conservation? Probably not (Poiani et al. 2000; The Native Conservancy 2000). Let's consider some "points of view" that other animals and plants might take and reflect on whether those should influence the inquiry process and the development of conservation guidelines.

Different Points of View on Time

As a human being, you have a unique viewpoint on time as well as space. You're active throughout all seasons for a number of years. Unconsciously or consciously, though, you might think and plan in terms of the "average" year. Most of your inquiries, management plans, thesis projects, contracts, or funded projects last for one to five years even if, after reflection, you extend results to much longer time spans. Your professional lifespan will last somewhat longer than a single project, I trust, but probably won't exceed four or five decades.

Perspectives of Organisms with Different Life Cycles

The human perspective on time somewhat resembles that of an ant or termite colony, which lives through many years and often experiences the entire seasonal cycle of each. Many insects and other invertebrates, though, experience a single season or less as adults, perhaps also appearing at a different season as larvae or nymphs with quite different lifestyles. Often insects entirely "avoid" unfavorable seasons or even years and truly experience only a set of climatic conditions that's much more restricted and predictable than what you perceive the climate to be. Some vertebrates essentially do the same. Even the most tropical of frogs often spends a substantial portion of the year inactive, underground, or otherwise insensitive to the day-to-day climate outside. A few frog populations may remain inactive for several years running. Thus, the points of view of such frogs or invertebrates involve only brief glimpses (by our standards) of certain seasons, perhaps only in certain years as well. How should you take this into account in designing inquiries, interpreting results (for example, the absence or superabundance of some invertebrate or amphibian species during a given month or season), and setting up conservation guidelines—or doing a thesis project?

Likewise, your thesis idea or conservation concern and the inquiry that results may involve shrubs or trees that produce flower crops year after year for decades or centuries. Whether such a plant succeeds each year at turning flowers into fruits, seeds, and seedlings may matter little if at all from any perspective except that of the animals and fungi feeding on fruits, seeds, and seedlings—and perhaps some of those animals don't "care" either (Herrera 1998). Perhaps a given individual succeeds in contributing large numbers of seedlings to the surrounding landscape only once or twice during its long lifetime. Your concern should be with that episode, not with the events of the average year. Perhaps the population as a whole enjoys significant recruitment (lots of new seedlings) only during infrequently occurring "mast" years when huge numbers of seeds overwhelm the animal seed predators. If you observe no recruitment for one or several years or other disturbing patterns in plant reproduction during a particular year, should you articulate a conservation concern, embark upon the management cycle of figure 2.4, and develop management guidelines intended to conserve that plant population? Not necessarily. Try taking the plants' point of view.

Perspectives of Especially Long-Lived Plants

It's tempting to assume that the native plants currently flourishing around you are "in tune" with the present-day physical and biological environments where they're found, and to develop your

Figure 6.1.
An alerce tree (*Fitzroya cupressoides*) in
Parque Nacional Los Alerces, southern Argentina
(Chubut Province). This individual is probably
at least two thousand years old.

management plans accordingly. After all, for the sake of convenience many other conservation professionals, ecologists, and evolutionary biologists make that assumption. Some plants alive and flourishing today began their lives under very different conditions, though. The alerce tree (*Fitzroya cupressoides*) of figure 6.1 landed on the soil as a seed, germinated, and began growing about the time the concept of natural history was just a gleam in the eye of Aristotle or Pliny. During the tree's life, southern South America has undergone many climate changes. Can you assume that the physical and biological conditions you measure today at the roots of this tree are identical to those under which it, and others nearby, germinated? Will you base conservation plans for the alerce population, from new seedlings on up, on the conditions under which mature alerces exist at present?

A two- or three-millennium-old alerce isn't even the most extreme case. In the western United States many of the clones of quaking aspen trees (*Populus tremuloides*) that dot mountain ranges (figure 6.2) began their careers at least ten thousand years ago. Each clone is technically a single plant individual. Researchers estimate that one particularly large clone has lived a million years or more (Mitton and Grant 1996). This plant has seen ice ages and warm wet interglacial periods come and go. Is it sensitive to the ephemeral successes and failures of this year, decade, or century? Might your own conservation concerns involve some plant individuals or whole populations that are "living relics"? If so, what will be your basis for developing guidelines for those beings whose perception of time is so much slower than your own?

Figure 6.2.
Clones of quaking aspens mixed
with conifers in the Rocky Mountains
(Colorado, USA). Each distinct shade
of aspen foliage signifies a different
individual (clone). Larger clones in
the photo could be ten thousand
years old or more.

Who Besides You Cares about the "Average" Climate?

From the point of view of most plants and animals (now including most conservation professionals), there's no such thing as an average year. Dry and wet seasons, cold and warm seasons, are not identical from one year to the next. The biota's responses to an inquiry's design factors under one year's climatic conditions do not necessarily predict responses under the next year's. As chapter 4 pointed out, inquiries lasting only a year or even two should be viewed quite skeptically if you're basing longer-term management guidelines on them.

Furthermore, extreme climatic events that occur only sporadically, such as those associated with the El Niño and La Niña phenomena, probably exert much more profound and long-lasting effects on populations, communities, and landscapes than do "normal" year-to-year fluctuations in climate. Such events may drastically alter the availability of resources to plants and animals, lead to booms or crashes in population densities of plants or animals, change temporarily the nature of species interactions, drive populations outside their usual geographic ranges, and divert the process of ecological succession on disturbed sites, leading to near-permanent changes in a site's vegetation and consequently in its animals. Even with the best of intentions, your inquiry, contract, or funded project that lasts one to five years is apt to miss the extreme event that really structures the ecological processes and species interactions in your landscape. Guidelines or management plans based strictly on your results may turn out to be inadequate when the next extreme event strikes. Will the time frame for any of the designs in chapter 4, even design 16, be certain of including the extreme climatic events that might well dictate the long-term response of fauna and flora to selective logging?

Footprints Left by Extreme Events of the Past?

Depending on their history, different biotas might have quite different perspectives on natural disasters that, to us, seem equivalent. The biota of the Galápagos Islands and the central South American coast, filtered through numerous El Niño episodes, would undoubtedly survive through future episodes if not further stressed by novel forms of human intervention and the invasion of exotic species. The coastal biota of Argentina, though, might be unable to survive a single severe El Niño–type episode simply because those sorts of events don't occur, or don't occur so dramatically, on that

Figure 6.3.
A Costa Rican landscape less than three years after Volcán Arenal's violent 1968 eruption. By the early 1980s this landscape once again supported robust secondary vegetation (and associated animals) except where converted for agriculture.

side of the continent. A volcanic eruption on the geologically calm Guyana Shield might have catastrophic, long-lasting effects on the local biota, whereas the local biota in some parts of geologically hyperactive Central America has survived and evolved through numerous such episodes (figure 6.3) and quickly exploits the altered landscapes that result from volcanism. Montane forests in the central Andes might be devastated by, and recover slowly from, a severe cyclonic storm, whereas montane forests on Puerto Rico can almost be said to thrive on the hurricanes that batter them with great frequency (Walker et al. 1991). How will your inquiries and conservation guidelines incorporate the unique point of view that your reserve's plants and animals have on extreme events?

Footprints Left by Past Climate Change

The whole globe, not just southern South America where the alerce trees live, has experienced dramatic climate changes over the geological past. Many episodes of rapid cooling, rapid warming, rapid wetting, and rapid drying have occurred over the past million years in particular. The most recent glacial episode ended quite abruptly with a rapid warming only around twelve thousand years ago, an event perhaps witnessed by the great-great-grandparents of the alerce in figure 6.1. These rapid fluctuations have kept animal and plant populations constantly scurrying about the continent trying to keep up. Not all have scurried at the same rate or in the same direction. Different plant and animal populations have retreated or advanced across the landscape (or up and down mountain slopes) at different rates. Few if any tightly integrated "ecological communities" have migrated back and forth across the landscape as units. Whether in the North Temperate, subtropical, tropical, or South Temperate zone, species assemblages have been reshuffled time and time again.[1] Have these events left any footprints on today's landscapes?

Many investigators feel that few local biota, temperate or tropical, have yet reached an equilibrium in species composition—if indeed they ever reached in the past, or could ever reach, such an equilibrium. At any given site some populations may slowly be going extinct because their ideal climatic conditions moved elsewhere long ago. Other laggard populations appropriate for the site's present-day climate, having retreated elsewhere during the last glacial episode, haven't yet managed to return. What does this have to do with management guidelines? *Even in the absence of any further climate change, invasion by exotic species, or changes caused directly by human activity, the species com-*

position of your protected area would undoubtedly change. For that reason alone you'd be foolish to set up conservation guidelines with the hope that by doing so you could maintain the present-day species composition indefinitely.

Trampling by Future Climate Change

Perhaps the most urgent large-scale conservation problem you confront concerns future, not past, climate change. No one can state precisely what's going to happen to world climates over the next few decades or centuries. Most researchers agree, though, that much of Latin America may soon experience an unprecedented temperature rise of several degrees Celsius and drastic changes in rainfall regime, not to mention increased levels of CO_2 that may dramatically affect plant growth and plant-animal interactions.[2] Most organisms living in flat areas will see their ideal climatic conditions migrating rapidly polar-ward. Most organisms living in mountainous areas will see their ideal climatic conditions migrating rapidly upslope. The only way populations will survive is to scurry about, tracking these rapid movements of their preferred climatic condition, more rapidly than ever before in their histories. Today's climate change won't just leave footprints on landscapes, it will trample them. Your best efforts may be insufficient to prevent some species from going extinct, even during your professional lifetime, within the protected area you manage. Perhaps you can stem the loss a bit, though, by considering the shape of your reserve and the diversity of habitats it encompasses.[3] Ponder the following questions:

1. If your reserve is in mountainous terrain and it's possible to expand protection in one direction but not all, should you try to expand (a) into new habitats along the same elevation belt or (b) upslope? What are some inquiries that might guide your choice?

2. If your reserve is on flat terrain, should you try to expand (a) the total area protected of each critical habitat already existing in the reserve or (b) the number of different habitats that are protected? What are some inquiries that might guide your choice?

3. Are you sure that the geographic sites where today's "biodiversity hot spots" (Mittermaier et al. 1998) occur—and upon which many local, regional, and global conservation initiatives are based—won't be tomorrow's "biodiversity hell" as temperatures shoot up and species either flee or go extinct? Have all proponents of hot-spot thinking adequately taken climate change into account?

Even if the earth experiences climate change that's less rapid or less intense than what most researchers fear, some warming and some shifts in global precipitation patterns will undoubtedly occur over the next century or less. Human conversion of the surrounding landscape (chapter 7) may limit the ability of plant and animal populations in your protected area to scurry about in response. Therefore, you must consider climate change when developing management guidelines for the foreseeable future. Or, if you tend to get depressed easily and you manage a very small reserve that's surrounded by pavement on flat terrain, perhaps you'd best try *not* to think about the future.

Footprints Left by Past Human Inhabitants

Climate fluctuations aren't the only changes that landscapes in the New World experienced during the few thousand years preceding the European Conquest. Do you truly believe that you're managing a "pristine" landscape, where pristine means "free of past human influence"? If so, you're almost certainly wrong. Wilderness unsullied by human touch is a nice ideal. In the real world, though, there's no such thing as a pristine landscape except perhaps in polar regions.[4] Since their arrival twelve thousand years ago or so, humans have trodden upon nearly every square meter of the Western Hemisphere's landscapes at one time or another. In some cases, footprints were few and light. In others, such as eastern North America, much of Mexico and Central America, and much of northern and central South America, footprints were many and heavy. Great numbers of indigenous peoples died as a direct or indirect effect of the European arrival. Their footprints have since faded somewhat even from the landscapes they once trampled most heavily, to the point where those landscapes may now appear pristine to the inexperienced eye. Please note that the term *footprints* is a metaphor—other authors might say effects, signals, artifacts, or impacts.

Clearly, the "lush tropical jungle" of southern Mexico and northern Central America is anything but pristine, having supported the vast Mayan civilization and others. Effects of past civilizations on South America's altiplano are especially conspicuous (Binford et al. 1997). Less obviously, millennia ago indigenous people cultivated maize over much of the Darién Isthmus of Panamá and Colombia (Bush and Colinvaux 1994), a region sometimes misnamed "the Western Hemisphere's last frontier of pristine rainforest." "Virgin rainforests" along the banks of the Amazon and its major tributaries, the icon of schoolchildren and environmental activists of North Temperate nations, apparently supported intensive agriculture and up to several million people until soon after the year 1492.[5]

Even landscapes with no extensive pre-Conquest civilizations felt the tread of humans (see Martin and Szuter 1999). In tropical wet or moist forests, seeds of today's towering canopy trees with edible or otherwise useful fruits may have germinated in human latrines or in "forest gardens."[6] Conversely, indigenous hunters before and since the Conquest may have interfered with seed dispersal by other large mammals, and altered the dynamics of many wet tropical forests, by thinning or extinguishing populations from some sites (Redford 1992). On a much larger scale, Paleoindians permanently altered the ecology of the vast Chaco regions of South America as well as other tropical or subtropical dry forests, first by apparently extinguishing a great many large mammal species (or at least witnessing their extinction) and then by setting frequent fires to smoke out the survivors.[7] If you haven't yet perceived any traces of past human footprints in your protected area, either you haven't dug a deep enough hole or you haven't talked with the local anthropologist.

As a professional concerned about conservation in the present and future, why should you worry about the footprints left by pre-Conquest humans? First, if you intend to restore part or all of the reserve to its "natural" state (see below), which set of past conditions will you choose to represent that state?[8] If you pick any date that's more recent than 12,000 B.P., you'll probably need to incorporate the footprints of one human group or another. If you pick a slightly earlier date to avoid dealing with human footprints, you're up against the Ice Age and, in many cases, a suite of extinct large mammals whose influence cannot be duplicated today. Second, should you take the footprints of past humans into account when proposing conservation guidelines that regulate use by present-day humans? It's quite possible that a landscape intensively used by pre-Conquest humans could rebound quickly from moderate intrusions by present-day humans, just as the biota of a hurricane-prone Caribbean island rebounds quickly from a hurricane, whereas a superficially similar landscape that was used only lightly by pre-Conquest humans might be sensitive to the slightest human intrusion in the present day. To my knowledge this possibility has never been investigated. What sorts of inquiries could you design on this subject, and what management alternatives might result?

Footprints Left by Recent Disturbances of Human or "Natural" Origin

Natural landscapes experience frequent disturbance events even when humans are absent. Branch-falls and treefalls open small to medium-sized gaps in forests. Severe windstorms and landslides open large gaps (figure 6.4). Fires and hurricanes or cyclones open still larger gaps. Far from conforming to earlier views of "climax" communities, forests and other habitats are mosaics of recently disturbed patches, patches disturbed some time ago and now partway through regeneration, and expanses by chance undisturbed for some time (figure 6.5). A habitat's biodiversity, its richness of ecological actors and acts, may depend in large part on this disturbance mosaic.[9] Some human activities, such as extensive agriculture or industrial forestry, create huge, catastrophic disturbances that overwhelm the rich, small-scale mosaic. Other human activities, though, involve a scale similar to that of non-human disturbance events and simply amplify the "natural" mosaic. Examples might include slash-and-burn agriculture at low intensities (see Andrade and Rubio-Torgler 1994) and subsistence forestry. Some recent designs for "natural forest management" consciously strive to mimic the scale and intensity of disturbances resulting from wind-caused treefalls (reviewed in Meffe and Carroll, 1997: 603–608).

Given the task of directing your protected area toward its "natural" state, then, should you strive to prevent all forms of disturbance in order to achieve "climax" primary vegetation throughout? Absolutely not. Should you design management guidelines that allow instead for "natural" levels of disturbance, that encourage directional change (regeneration) of the biota on disturbed sites? Absolutely.

Now consider regeneration from another practical perspective. What if you observe some changes in the vegetation within parts of the reserve currently used by local people? Should you blame the present-day intruders for those changes? Not without better evidence. For example, sup-

Figure 6.4.
Vegetation regenerating on the scar left by a landslide (Parque Nacional Podocarpus, Zamora-Chinchipe Province, Ecuador).

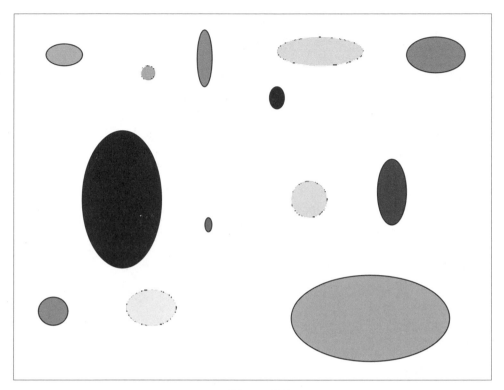

Figure 6.5.
Schematic diagram of a natural disturbance mosaic—for example, a forested landscape. The darkest patches represent the most recently disturbed sites. As regeneration takes place, these fade into the physical conditions and species assemblage characteristic of the regions not disturbed for some time (light-colored background). Of course, meanwhile new disturbance events occur, so at any one time the landscape presents a mosaic of patches in all the different stages.

pose you manage a reserve that was put under absolute protection twenty years ago. Before that time, outer reaches of the forest experienced some slash-and-burn agriculture and scattered pit-sawing of valuable timber trees. Five years ago, you grudgingly agreed to reopen the outer reaches of the reserve to low-level use. Under the new program, surrounding communities harvest nontimber forest products such as vines for basketry and herbs for medicine. Now your management concern is whether to discontinue, maintain, or expand the harvesting program. You've noticed some recent changes in the vegetation of the zone where harvesting takes place. You're tempted to blame the current harvesting program for these changes and to argue for discontinuing the program. Is that fair? How might you design an inquiry capable of distinguishing between the current effects of the harvesting program and effects of regeneration from past disturbances?

What's "Natural"?

By now I hope it's clear that the distinction between "natural" landscapes and landscapes "sullied by humans," between events labeled "natural" or "unnatural," is unfair. Instead, a continuum exists from those few landscapes (if any) with no significant footprints ever left by human beings, to today's most highly modified agricultural, industrial, overgrazed, or urban landscapes. For better or worse, humans are a part of nature (Soulé and Lease 1995; Hunter 1996). Our influences are inextricably entwined in the history and current status of nearly every present-day landscape. While developing management guidelines, how will you recognize this continuum? Should you assume that all "westernized" approaches to landscapes are noxious and to be avoided, while all indigenous approaches to landscapes are beneficial? No, that's often another false dichotomy. The romantic myth of the "noble savage" who lives in harmony with the environment is often just another dream of political and environmental activists in the North Temperate zone and in the large cities of your country.[10] Westernized approaches to conservation in Latin America aren't automatically bad—after all, the conservation ethic and your own conservation guidelines are westernized approaches—and "native" approaches to landscapes aren't automatically good.

Likewise, what do the labels "native species" and "exotic species" mean to you, and how should they influence your management guidelines?[11] Invasions of exotic plants and animals constitute a primary conservation concern globally and locally. Managers expend considerable effort and resources to eradicate exotic invaders. Does the label "exotic" always mean "bad" from an objective conservation standpoint? Often, but not always. After all, in the historic sense many or most of your reserve's species are exotics, having arrived there at some time during the past twelve thousand years. Several common medicinal herbs widely used in the South American countryside actually arrived with early Europeans. Some avowedly "natural" populations of plants and animals may be results of intentional releases further in the past. Certainly, many edible wild plants, and perhaps some edible wild vertebrates as well, owe their present-day geographic distributions to conscious transport by pre-Conquest peoples. Some present-day populations of "native" birds may have originated as pre-Conquest releases for purely aesthetic reasons (Haemig 1978, 1979).

Furthermore, just because an exotic species is aesthetically disagreeable doesn't necessarily mean it has negative effects on local biota. The rosa mosqueta *(Rosa eglanteria)*, an aggressive and thorny invader of protected areas in southern Argentina (Damascos and Gallopin 1992), may actually enhance the regeneration of natural forests at some places by protecting many native tree seedlings from grazing by other exotic species: cows, sheep, and European red deer (J. Salguero, personal communication). Finally, a given exotic species may affect some landscapes much more severely than others. By any measure feral goats, other exotic mammals, and several species of nonnative plants are currently affecting the native biota of the Galápagos Islands to an alarming extent.[12] Except for goats, though, few of those exotics have nearly such severe effects on the mainland. Even feral goats rarely affect mainland habitats so severely that conservationists must plead for international expeditions to hunt them out (MacFarland 1991).

I certainly do not intend to downplay the extraordinary threats posed by many exotic species invasions and the homogenization of the world's biota (see Putz 1998; Richardson 1998). I do intend to convince you, though, of the danger in setting up dichotomies and assigning labels having connotations of "good" and "bad" that direct your management decisions from then on. In many tropical and subtropical regions, including the state of Florida in the United States, exotic species tend to affect protected areas most seriously on islands and peninsulas (Simberloff, Schmitz, and Brown 1997; Fritts and Rodda 1998). In continental regions, wet tropical landscapes seem to suffer less from exotic invaders than do semi-arid or temperate landscapes (e.g., Veblen et al. 1992). Should your guidelines require you to fear all nonnative species and eradicate them from the landscape whatever the cost? Should you assume automatically that a given species labeled elsewhere as "invasive" or "noxious" constitutes a serious threat to the integrity of your own protected area, a threat that must be countered? How might you design inquiries to evaluate the threat that a given exotic species, plant or animal, poses to your local landscape?

Different Points of View on Space

Do all the inhabitants of the landscape where you work share your 1.6-m-high and 100-m-wide point of view? Of course not. For "proof," turn to the exercise in box 10.1 (page 153). A mite, an ant, or a mouse operates on a scale very much smaller than yourself. The mite perceives a sand grain as a giant boulder, a dead grass stem as a huge log that must be clambered over. At the other extreme, the eagle or Andean condor soaring overhead perceives a landscape scale much greater than your own and pays no attention to sand grains or grass stems. Other distinctions are more subtle, however.

As a practical way to think of spatial scale, let's say that the *ecological neighborhood* of a plant or animal consists of the other individuals, of the same or different species, with which it interacts, and the area they cover. Below I'll list a few of the numerous sorts of ecological neighborhoods that undoubtedly exist among your reserve's biota and suggest ways to incorporate these diverse viewpoints into management guidelines (see also Schwartz 1999; Poiani et al. 2000; The Nature Conservancy 2000).

A Flowering Shrub

The extent of a plant's ecological neighborhood depends on which facet of its life you consider. For example, the "floral neighborhood" of a flowering shrub in the understory of a tropical cloud forest, or any other habitat, consists of the other flowering plants nearby and sharing the same animal pollinators, such that all might be visited during a single foraging bout by one pollinator individual (Feinsinger et al. 1986; Feinsinger, Tiebout, and Young 1991). This neighborhood often has a radius of a few to a few dozen meters. The radius of the shrub's "pest neighborhood," consisting of the plants (usually conspecifics) with which it shares insect pests, may be somewhat smaller or slightly larger, depending on the scale at which the pests themselves operate. The shrub's "resource competition neighborhood" might include only the closest neighboring plants and involve a much smaller area. By any definition, then, a plant's ecological neighborhood has a radius of tens of meters, or hundreds at the very most. It appears that you might indeed be able to treat plants as if they shared your own point of view on spatial scale.

A Hummingbird

The hummingbird that pollinates some understory shrubs and by its actions creates the floral neighborhood may not have a much larger ecological neighborhood itself—for the moment at least. Some hummingbirds with fairly short bills may forage at a few profusely flowering trees or canopy epiphytes for most of the day or may traverse a hectare or two of less flower-rich forest understory. Hermit hummingbirds and others with long bills, though, often repeat foraging routes several kilometers long many times each day. Over the course of the year the hermit's trapline route will probably shift as plants come into or go out of flower but remain in the same region of forest. The bird's ecological neighborhood may thus encompass one to several square kilometers over the year, an area probably much larger than your own concept of the "ecological community."

Some individuals of some short-billed hummingbird species may also remain in a certain region of forest throughout the year, resulting in an ecological neighborhood about the same size as that of their longer-billed relatives. Other individuals of the same species, though, and entire populations of many short-billed hummingbird species migrate seasonally (a) among neighboring habitats; (b) horizontally for tens of kilometers, exploiting nectar and arthropod resources in very different vegetation types; (c) vertically, up and down mountainsides, or (d) latitudinally, for hundreds or thousands of kilometers.[13] Some individuals of one atypical genus of hermit hummingbirds, the sicklebills (*Eutoxeres*), may change elevation by thousands of meters each day as they forage up and down tropical mountainsides. The ecological neighborhoods of hummingbirds that migrate don't resemble in the slightest the human spatial scale or even the scale of the landscape in which a protected area sits. If your goal involved conserving hummingbirds or another group with highly variable sizes of ecological neighborhoods, how relevant would be an inquiry, or a management plan, carried out even at the scale of your entire protected area?

A Resplendent Quetzal

Many fruit-eating birds, and some fruit-eating mammals as well, also circulate among different habitats over the year (Levey and Stiles 1992; van Shaik, Terborgh, and Wright 1993). One charismatic fruit-eating bird, the resplendent quetzal of Central America, characterizes lush cloud forests—during part of the year. At other seasons, quetzals move to very different, drier forests at lower elevations. In northwest Costa Rica their ecological neighborhoods span not only different forest types but also different categories of land use, from large protected areas in cloud forest to small, vulnerable forest patches downslope, nearly always on private lands and surrounded by fields or cattle pasture. Conserving quetzal populations is a very real concern for personnel of protected areas in cloud forest—and a practical concern too, for quetzal watchers bring many dollars, francs, and marks. Guidelines that involved only the spatial scale normally investigated by field ecologists—for example, a patch of cloud forest—would obviously be inadequate. Instead, conservation initiatives for the quetzal must address the scale of its ecological neighborhood (Powell and Bjork 1995).

A Migratory Shorebird

The prize for largest ecological neighborhood among birds probably goes to certain migratory shorebirds that breed in the Arctic and spend much of the nonbreeding season on coasts of southern South America, stopping at ocean beaches or altiplano lakeshores en route. A conservation scientist working in a protected area concerned with shorebird conservation recognizes that the necessary initiatives must involve an immensely larger scale than his or her own management guidelines can address.

Other Animals

I selected a few bird examples, and one plant example, to illustrate the great variety of points of view on space and the challenge of incorporating these various points of view into management guidelines (see Thiollay 1989). I could just as easily have chosen another animal group. For instance, ecological neighborhoods of some butterflies and moths conform to our own biased one-hectare-scale perspective on the ecological community, but some hawkmoths forage up and down steep tropical mountainsides each night or move between tropical dry and wet forests. Some butterflies engage in seasonal migrations, which may be spectacular and world famous (Brower 1997). Likewise, some nonflying mammals may perceive the landscape at our scale or even a smaller one. As you know, though, some carnivores and large herbivores such as tapirs have extensive home ranges, while even in the Western Hemisphere (let alone Africa) other large herbivores may migrate seasonally among habitats. Among bats, the range of ecological neighborhoods of different species, sexes, or individuals may almost equal the variation among hummingbirds. While some bats are quite sedentary, others forage nightly over many square kilometers and diverse habitats, and still others may migrate seasonally. If we expand our perspective to the oceans, then the ecological neighborhoods of some mammals (whales) and reptiles (sea turtles) may rival those of shorebirds. If bat, butterfly, hawkmoth, whale, or sea turtle conservation is an issue, how will your conservation guidelines address the animals' points of view?

The Design and Interpretation of Inquiries

In chapter 4, we worried about the possible effect of selective logging on forest birds, mammals, and frogs. All designs involved comparisons between response units that were either SL tracts (in design 13, tracts yet to be logged) and UL or control tracts of similar size, whatever the scale and number of evaluation units per tract. A typical logging parcel encompasses a few tens of hectares at most. If we now consider the animals' points of view on spatial scale, can the inquiry still address the management concern? For frogs, a qualified yes. The ecological neighborhood of some forest frogs probably encompasses only a small fraction of the SL tract or comparable UL region. Therefore, SL tracts may not impinge on the ecological neighborhoods of frogs in UL control sites nearby. The ecological neighborhoods of a few forest birds may also conform to the scale of SL tracts and UL controls, but those of many others will encompass far greater areas that might easily include both SL tracts and UL sites at once. If such birds react either positively or negatively to the holes that selective logging punches in parts of their neighborhoods, then their actions in the UL portions of neighborhoods will change as well. In consequence, even if confined to a few tracts in a large, otherwise unlogged reserve, selective logging could well alter the nature of bird assemblages not only within SL parcels but also within UL regions quite some distance away. A study design using those UL regions as controls may be biased (see also Crome 1997).

Thus, even the most rigorous designs for the inquiry in chapter 4 might fail to test objectively the effect of selective logging on forest birds at the scale appropriate to birds themselves. From many birds' points of view, individual SL tracts or equal-sized UL regions aren't truly independent response units. Perhaps the only truly independent response unit is an entire reserve in which some selective logging either does or doesn't take place. As the question stands in this case, you're evaluating only the one reserve, so in essence you have only one response unit of one level (some selective logging) of the design factor and none of the other level (control). To truly evaluate the effect of selective logging on forest bird assemblages, you may need to replicate reserves instead of replicating SL and UL tracts within the one reserve.

Perhaps the ecological neighborhoods of most small mammals conform to the scale of logging parcels, and you're safe with the original inquiry. Possibly, though, some species or individuals will have larger-scale points of view such that their use of UL regions reflects their response, positive or negative, to SL tracts nearby. Without checking into the mammals' natural history, can you assume that mammal assemblages in the UL controls are the same as they would be if no selective logging had ever taken place nearby?

In short, if your conservation concern specifically involves birds, small mammals, or other animal groups with varied points of view on space, consider the scale of the design factor from their point of view, not yours. No matter how clever and statistically sound your design in theory, if you neglect to take points of view into account, the results, reflections, and applications you generate might be erroneous. As the final section of this chapter stresses, don't be discouraged by this real complication of natural history—just be careful. For practice, go to box 6.1.

Box 6.1. Incorporating Diverse Points of View into Your Design

Return to the design you proposed during the exercise of box 4.2. If the question behind your design specifically involved species other than humans, did you take their viewpoints on time and space into account? If not, how would you redesign the inquiry to do so? If you now find it impossible to design an inquiry that complies not only with the viewpoints of its subjects (this chapter) but also with the requirements of design and statistical inference (chapters 4, 5), how might you rephrase the question or redirect it to a different group of organisms?

Different Points of View on the Site Itself

Let's leave birds and wandering mammals aside for the moment and visit a population of plants, or of quite sedentary animals, that spreads over a certain expanse of the landscape. Would you expect the population's dynamics to be uniform from one end to another of that expanse? Might reproduction be more successful in some spots than in others? Might adult individuals survive better in some spots than in others? That is, might birth rates and death rates vary from point to point across the population's range?

Now let's assume that on average, taking into account its entire range, the population is stable. That is, as a whole it's neither increasing nor decreasing. In more precise terms, on average births exactly balance deaths, and the *finite rate of population growth* (λ) is 1.0. Following the reasoning in the previous paragraph, though, at a finer scale the value λ will be certain to vary from point to point, from > 1.0 in regions where births exceed deaths to < 1.0 in regions where deaths exceed births. Does this make sense to you? This simple proposal, far from being esoteric, actually ties together a great many concepts in conservation ecology, provides some novel insights into biodiversity conservation, and has great significance for management decisions.

Sources and Sinks in the Simplest Case

Let's define those points where $\lambda > 1$ as *sources,* their excess of births over deaths generating "extra" individuals over time, and those points where $\lambda < 1$ as *sinks,* where the population is maintained steady only if the spillover of "extra" individuals from nearby sources counteracts the local excess of deaths over births (Dias 1996; Meffe and Carroll 1997: 211–213). If we could map the source-sink "topography" of a given population, the result might resemble figure 6.6, with source patches scattered among a background of sink regions. Some, much, or nearly all of the range of a plant or animal population might actually consist of sink habitat, incapable of sustaining the population over the long run without input from the scattered source patches. Obviously, the model applies to less sedentary species as well. After all, even migratory hummingbirds and shorebirds settle down to breed somewhere, sometime. Within that circumscribed breeding range there's almost certain to be a source-sink topography like that of figure 6.6.

Can you really map a population's sources and sinks or recognize differences between the individuals found in each? Probably not. The individual plants or animals in sink areas will interact with each other and with other species, affect and be affected by their ecological and physical neighborhoods, mate, reproduce successfully, and go through life just as cheerfully as their counterparts that

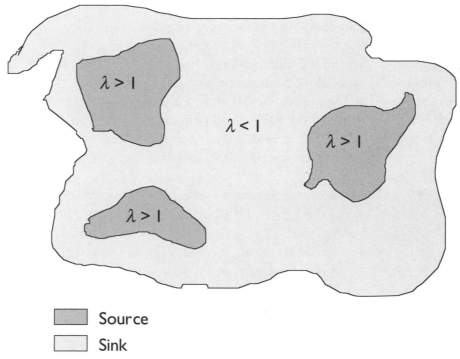

Figure 6.6.
Sources and sinks within the range of an animal or plant population. Overall the population may be neither shrinking nor growing ($\lambda = 1$), but in sink regions deaths exceed births ($\lambda < 1$), while in source regions births exceed deaths ($\lambda > 1$). In this case, most of the range happens to consist of sink habitat, where the population could not sustain itself were it not augmented by overflow from the source regions.

happen to live in sources. Population density in sinks may equal or exceed that in sources. The source-sink topography isn't a visible feature, or a cause of the variation in λ. Rather, it's a long-term consequence of that variation, of the invisible, often tiny, differences between birth and death rates. The only way to map the source-sink topography of a real population would be to follow birth and death rates throughout its range over a long period of time—a *very* long period of time if you're concerned about alerces, quaking aspens, ocelots, or tapirs. And any source-sink topography will certainly fluctuate over time. I doubt that you could design an inquiry on the subject of sources and sinks. So of what use is the concept?

Source-Sink Topographies, Time, and Space

First, the source-sink model may be the most useful way to view responses of populations and entire species assemblages to (a) the disturbance mosaic and (b) climate change. As to the disturbance mosaic, for many plant and animal populations in the protected area you manage, regions undisturbed for some time may be sinks, while small, resource-rich, recently disturbed or regenerating patches provide the only sources. Obviously, the map of sources and sinks will change over time as previously disturbed sites return to "mature" conditions and new disturbances pockmark the landscape. How might this perspective influence the inquiries you undertake and the conservation guidelines that result? On a larger scale, sources and sinks migrate across the landscape as climate changes. What was formerly a source will now be a sink, and what was formerly a sink may now become a source—if the population can get to it. Does this suggest novel ways to manage protected areas, or the semi-natural matrix outside them, in the face of global change?

Now consider instead the population's source-sink topography at one point in time, but this time with respect to a gradient of habitat conditions (figure 6.7) instead of geographic coordinates (figure 6.6). For example, proceeding from the driest point to the wettest point within the population's range, or the lowest to the highest elevation, you'd tend to encounter its source patches near the middle with sinks stretching to the habitats on either side. Such environmental gradients characterize real landscapes, of course. Therefore, a given population's source zones will likely center in one climate zone, vegetation type, or microhabitat, spewing their overproduction into the greedy sinks on either side (figure 6.8).

Of course, different species will have different source-sink topographies. A habitat that's source for some populations will be sink for others. Of all the species apparently thriving in a given microhabitat, vegetation type, protected area, or climate zone, some will actually be in sources, others in sinks. The steeper the environmental gradient, the more closely packed the habitats along it, the closer any given point will be to other habitats, and the more each point will receive spillover populations from nearby sources—that is, the more sink populations each point will sustain. In consequence, the more closely packed the habitats, the greater will be the proportion of a given site's species assemblage that actually involves sink, not source, populations.

How does this scenario affect your life as a conservation professional? Some, many, or most of the species in a protected area, even those species that appear to be in the bloom of health, might actually be in sinks, not sources. Their presence depends, or has depended, on spillover from sources nearby. *If cut off from those sources, many will go extinct even without any influence of climate change.* The reserve's present species assemblage may well be unsustainable regardless of management

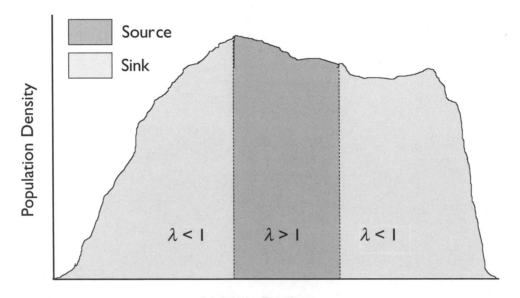

Figure 6.7.
One possible distribution of a population's sources and sinks along a habitat gradient.
Although at present the population as a whole occupies a broad range along that gradient,
source habitats span a much narrower range.

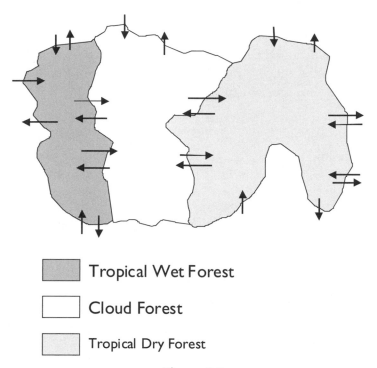

Tropical Wet Forest

Cloud Forest

Tropical Dry Forest

Figure 6.8.
Sources and sinks at the level of a landscape and its entire biota—for example, the reserve that served us in chapter 4. Each plant or animal species with its source in one habitat might spew individuals into neighboring habitats, sustaining sink populations that seem to be robust but would decline to extinction were the source cut off. Thus, the species assemblage of any one particular habitat includes "source" species that truly pertain to that habitat plus any number of "sink" species that really don't belong there, whose persistence depends on the subsidy from their own sources elsewhere. Of course, sources and sinks would actually occur at a much finer level than just the reserve's three major habitat classes shown here. For example, the varied topography, elevation, temperature, and precipitation of the cloud forest alone would create a complex patchwork of sources and sinks for any and all species. Note that the source-sink process may involve not only "natural" habitats within the reserve but also the interchange between those habitats and the converted landscapes surrounding the reserve, the theme of chapter 7.

efforts, for reasons of source-sink topography alone. This problem may be particularly acute if your protected area is in a region of high biodiversity, for, by circular reasoning, a great deal of that biodiversity may result from just such source-sink dynamics.[14] The eastern foothills of the tropical Andes or some regions' rapid transitions between tropical wet and dry forests might exemplify this condition. Are some serious concerns, ideas for alternative conservation guidelines, and ideas for inquiries beginning to stir in your mind as a result of source-sink thinking? Perhaps chapter 7 will stimulate this thought process further. Source-sink topographies and dynamics don't stop short at the legal border of your reserve.[15]

What to Do?

1. Listen to a recording of the Argentine artist Mercedes Sosa singing "Cambia, Todo Cambia" ("Everything Changes"). Don't ever forget that.

2. Generalize with great caution. Recognize once again that even the best-designed inquiry (chapter 4) and the most rigorous statistical inference (chapter 5) necessarily involve restricted scales in space and time. Indeed, given that natural landscapes don't comply well with the "fundamental rule of statistical inference," in field situations the entire statistical approach must be applied and interpreted with special care.

3. Take the point of view of the species in question. Don't assume that the individuals, species, or species assemblages of conservation concern share your human point of view on space, time, change, or long-term suitability of the site. Rely on your natural history knowledge to think like a tree, a weed, a hawkmoth or catfish or frog or earthworm or tapir or quetzal.

4. Recognize the diversity of ecological neighborhoods that exists among species and even among individuals within a species. Recognize that the scale of neighborhoods may be smaller, about the same as, or much larger than the scale of the design factor in your inquiry. Each case will call for quite a different interpretation. As in the selective logging example, sometimes the two scales will be so dissimilar that you can't answer the question objectively. If that makes you uncomfortable, turn to chapter 8.

5. Set down the statistics text and go get your boots dirty (Noss 1996). Yes, the realities of natural history complicate design, analysis, the development of management guidelines, and your life as a conservation professional or field ecologist. The better you know the natural history of your reserve and its denizens, though, the easier you'll find it to consider their divergent points of view as you propose questions, answer them with appropriately designed inquiries, and develop management guidelines that will truly enhance conservation.

CHAPTER 7

Contents and Context: Taking the Whole Landscape into Accoun

No park is an island.

—D. H. Janzen (1983)

We humans sculpture landscapes so that they conform to our own ecological neighborho thereby imposing our particular point of view on other animals and plants. For example, rural la scapes often display a complex mosaic of a few remnants of the original vegetation dispersed amo patches with different land use, each patch about 0.1–3.0 hectares in area except for large livest pastures (figure 2.3). Large expanses of the original vegetation not yet reached by the agricultu frontier may end up as protected areas. Of course, a protected area's legal boundary might a include a number of the previously farmed or logged patches. Conversely, substantial portions of continuous, original vegetation might remain outside that boundary. On occasion, though, the le boundary corresponds roughly or even precisely (figure 7.1) to the physical border betwe the remnant original vegetation and the surrounding mosaic, aptly described by the phrase *sem natural matrix* (Brown, Curtin, and Braithwaite 2001).

Traditionally, the manager of a protected area is charged with conserving its *contents,* the hab tats and biota within the legal boundary. Many conservation strategies have treated the protecte area's *context,* the semi-natural matrix outside, as a buffer zone (Wells and Brandon 1993) that serv to insulate the reserve from more seriously disturbed landscapes some distance away. Other than i usefulness as a buffer zone, should those conservation professionals affiliated with the protected are ignore the semi-natural matrix, treating the protected area as if its contents were sealed off from

Figure 7.1.
A protected area with a very sharp, distinct edge: Bwindi
Impenetrable National Park (Uganda).

intercourse with the outside world? Of course not. Let's examine a protected area's contents and context from the perspective of chapter 6 and from that of *landscape ecology*.[1] Note that this examination applies not only to large, protected areas but also to those smaller remnants of original vegetation where a great number of ecologists, social scientists, and conservation professionals often work (Schelhas and Greenberg 1996; Laurance and Bierregaard 1997).

The Protected Area from the Landscape Point of View

No matter how large a protected area or fragment, no matter how many different habitat types it contains, it's still just one element of a larger landscape. The protected area might seem to be quite self-contained, but like any other landscape element (for example, a lake, a tree plantation, a pasture, a field of crops, or the alpine zone of a mountaintop) it experiences some interchange and interaction with its surroundings. The nature of those interchanges and interactions depends on a great many features that include even such details as the direction of the prevailing wind or the exact geometry of the border between reserve and semi-natural matrix (Forman 1997).

Of course, the protected area and the semi-natural matrix shouldn't ever be viewed as two distinct, homogeneous classes of habitat. Almost always, "reserve" and "not-reserve" are entirely distinct only in the political sense. The not-reserve, the semi-natural matrix, consists of a wide range of habitats and species assemblages. If the landscape had been originally forested, then today's continuum of land uses and habitat types among the patches making up the semi-natural matrix might include carefully tended garden plots; pesticide-drenched crop parcels or plantations; improved cattle pastures planted with exotic grasses; carefully tended fruit orchards; carefully tended plantations of exotic trees such as eucalypts or pines; less carefully tended plantations with a robust understory

of native shrubs and saplings; traditional coffee or cacao plantations with native and exotic shade trees; fruit orchards with a diverse herbaceous understory; subsistence crops or unimproved pasture with a diverse weed flora and insect fauna; abandoned fields or pastures now sporting many plant pioneers of local and exotic origin; woodlots or other forest remnants still supporting some canopy-height native trees; diverse secondary shrublands and forests of all types, including tracts scarcely distinguishable from the forests protected within your reserve; and remnant patches of the original vegetation itself. If the native vegetation is something other than forest, the semi-natural matrix surrounding a remnant or reserve will still display a wide range of habitats that have experienced various degrees and kinds of conversion by local people. What useful points of view exist on the relations between these diverse surroundings and your protected area?

Have You Looked at Edges from "Both Sides, Now"?[2]

Let's begin with the simplest possible case: a quite small, isolated patch of the landscape's original vegetation that's now distinct from the converted lands that surround it. Such patches are the end result of the process known as *habitat fragmentation,* in which advancing agricultural frontiers or other forms of landscape conversion break up the formerly continuous natural vegetation and typically leave remnants in the less accessible or less cultivable sites.[3] Previous to fragmentation, ecological neighborhoods of the site's plant and animal inhabitants could extend freely in any direction. Today, though, the habitats next door to the fragment are very different from those within it, and many individuals' neighborhoods might stop abruptly at the fragment's edge. What happens next? Let's apply the source-sink model and see.

First, if you were flying in an airplane high overhead, the edge would appear as a sharp, absolute boundary between the reserve's contents and context (figures 7.1, 7.2a). Even close up, the edge remains an absolute barrier—from the viewpoint of those populations that suddenly "discover" they'd been living in a sink all along, subsidized by overflow from nearby sources that have just vanished. With no further input from such sources, as chapter 6 points out, these populations will wait in vain to be rescued from rapid or painfully slow extinction. Barring climate change, the only original species likely to persist indefinitely in the fragment of original habitat, no matter how large, will be those that by chance had sources within its borders when the borders were created.

From another viewpoint, though, the edge between fragment and surroundings is quite porous (figure 7.2b). Unless surrounded by asphalt pavement, the fragment is now accessible to new source populations, for the semi-natural matrix supports its own, different set of plant and animal populations that may spill across the boundary. These may encounter only sink conditions within the fragment. Nevertheless, because they're subsidized constantly from outside, these colonists may reach high population densities and exert quite an influence on subsequent events within the fragment (Janzen 1983). Colonists might be quite innocuous or they might be aggressive competitors, pests, pathogens, parasites, or predators of the remnant's original species.

The invasion of a fragment by species that thrive in the semi-natural matrix, natives of the landscape or exotics, constitutes one class of what are termed *edge effects.*[4] Physical edge effects may also infiltrate the fragment. For example, the raw edge of a forest fragment surrounded by newly cleared land will be exposed to open air for years or decades until growing vines, shrubs, and saplings seal it off. As a result, the outer reaches of the forest may experience greater variability than before in temperature, atmospheric humidity, soil moisture, and wind speed, all of which will affect the plants and animals living there. Edges of fragments are also vulnerable to pesticides and fertilizers that fil-

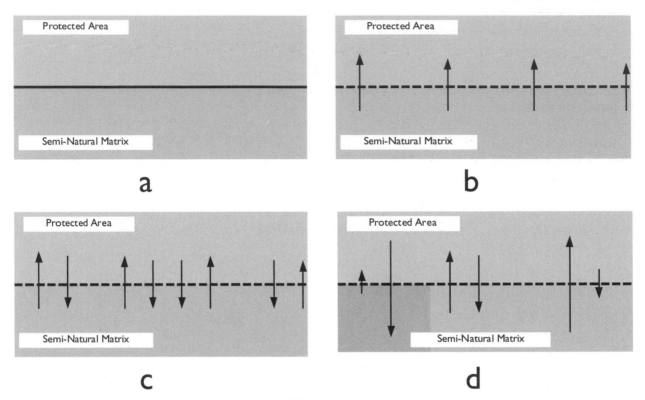

Figure 7.2.
Four different perspectives on the border between a tract of original vegetation and the semi-natural matrix (Brown, Curtin, and Braithwaite 2001), the converted or partially converted landscape alongside. (a) The "parks as islands" perspective: the boundary is an absolute barrier, particularly to the tract's surviving species. (b) The classic "edge effects" perspective: rather than being sealed, the border is permeable to physical influences (often inimical to the remnant tract's surviving populations and ecological interactions) and new species (often noxious) from outside. (c) The perspective of "both sides, now": the border is permeable to influences, interactions, and interchanges passing in both directions. (d) The "both sides, now" perspective, taking into account variation in the nature of the semi-natural matrix.

ter in from nearby farmlands. Whether physical or biological, edge effects may penetrate only a few meters or may reach several hundreds of meters into the fragment. A small fragment may experience edge effects throughout its area, inducing rapid changes in physical conditions and species composition from its preisolation state. Larger fragments may retain a reasonably large core area free from edge effects, while in reserves of hundreds of thousands of hectares, edge effects will involve a tiny fraction of the whole and may cause little concern.

There's yet another perspective on edges. Conservation concerns have traditionally focused on the fate of the protected area, not that of its surroundings. Consequently, most theoretical and practical work on edge effects takes note only of the arrows in figure 7.2b and concentrates on edge effects from the outside in. Don't edges have two sides, though? Might the physical environment of the remnant also penetrate some distance into the semi-natural matrix (figure 7.2c)? Might animal and plant populations residing in the fragment also spew colonists into the semi-natural matrix (Gascon et al. 1999; Griffith 2000) even if the matrix on its own could never sustain viable populations of those species?

The metaphorical pores in the border between your protected area and its surroundings almost

Figure 7.3.
Habitat shreds, common in many landscapes throughout the world (here, Tucumán Province, Argentina). The conservation concerns and conservation potential of shreds may be quite different from those of isolated habitat fragments.

certainly work both ways, not just one. Furthermore, the nature of the interchange between contents and context probably isn't uniform all along the edge (Mesquita, Delamônica, and Laurance 1999); surely it varies with changes in the nature of the semi-natural matrix alongside (figure 7.2d). Surely your protected area's contents exert a stronger, or different, influence on some types of semi-natural matrix than others, while some kinds of semi-natural matrix exert a stronger, or different, influence than others on the protected area's contents (McIntyre and Hobbs 1999).

How important is it to consider "both sides, now" from the perspective of figure 7.2d, if you're working with the protected area or with the semi-natural matrix itself? I'll propose that it's critically important in every case. In Latin America and elsewhere in the Western Hemisphere, how many inquiries, whether undertaken by managers or basic ecologists, have dealt with the concept presented in that figure? To my knowledge, only a few (e.g., Estades and Temple 1999; Gascon et al. 1999; Mesquita, Delamônica, and Laurance 1999), although numerous studies have addressed the one-sided model of figure 7.2b. What serious conservation concerns or intriguing study ideas spring up, what questions might you ask, what feasible inquiries might you design by staring long and hard at figure 7.2d?

The "both sides, now" perspective extends far beyond the case of a small isolated fragment. For example, habitat conversion doesn't necessarily isolate the remnant vegetation in disconnected fragments. Often it results in long, narrow strips of remnant vegetation that are still connected to one another or to large expanses of the same habitat (figure 7.3). The large expanses might, in fact, be protected areas (de Lima and Gascon 1999). These habitat *shreds* (Feinsinger 1997) are essentially "all edge." Incoming edge effects pervade the shred from one end to another, but outgoing edge effects might profoundly influence the semi-natural matrix into which the shred penetrates.

No matter what its shape and size, any expanse of remnant vegetation, whether protected or not, surely experiences and exerts edge effects all along its border. The nature of those effects should interest not only the conservation professional, land manager, and basic ecologist but also *the rural communities in the semi-natural matrix.* From the point of view of local inhabitants, the edge effects exerted by a small, privately owned fragment or an extensive reserve could be beneficial, detrimental, or both at once. Effects on croplands or pastures near the fragment might include (a) stabilizing soil moisture, stabilizing atmospheric humidity, and ameliorating drought; (b) ameliorating temperature extremes; (c) providing a windbreak; or (d) increasing the natural biological control of crop pests, through predators and parasites emerging from the reserve (see Power 1996). On the other hand, the reserve's presence might also increase crop losses to fungi, other pathogens, insect pests emerging from the reserve, and crop pests as large and abusive as spectacled bears in South America and elephants in Africa (Naughton-Treves 1998); or it might decrease crop production because plant growth is slowed by shade, or because flowering and fruiting rates of crop plants decline. Do these possibilities suggest to you some management concerns, questions, and inquiries? Might the community itself be anxious to know answers to those questions or to others it might have about the effects of forest fragments? (See chapter 10.)

The Semi-Natural Matrix as a Conservation Entity in Itself

Some conservation professionals continue to perceive everything protected by the reserve as "good" and everything outside as "bad." This perception results from the politically necessary but biologically unfortunate dichotomy between "reserve" and "not-reserve" combined with a second biologically unfortunate dichotomy mentioned in chapter 6, that between "pristine" landscapes and landscapes "sullied by humans." As a result of these labels, the semi-natural matrix often inspires loathing as a source of noxious invaders or pernicious physical influences. At best it's merely tolerated, as the buffer zone.

Consider again, though, the vast array of habitats and species assemblages that a semi-natural matrix might include. True, improved cattle pastures, cotton fields, or plantations of African oil palms might hold little to interest the conservation professional. Other sites, though, might support many species native to the landscape, even species (and ecological processes) of great conservation concern (Gascon et al. 1999; McIntyre and Hobbs 1999; Brown, Curtin, and Braithwaite 2001). Some patches, or the matrix as a whole, might even support a reasonable facsimile of the species assemblages and ecological processes of the original habitat. For example, the diverse tree overstory of traditional coffee plantations often supports a rich vertebrate and invertebrate fauna (Perfecto et al. 1996; Moguel and Toledo 1999). "Kitchen gardens" may be havens of local biodiversity (Gómez-Pompa and Kaus 1992; Steinberg 1998). Native forests managed for timber production may sustain bird assemblages nearly as rich as those of unlogged forests (Thiollay 1995; Griffith 2000). Small fragments of original vegetation scattered about the semi-natural matrix, scarcely given a second glance by many conservation professionals, may actually hold many native species, provide sources for regeneration of the converted lands that surround them, and hold great practical and even religious significance for local communities.[5] On a larger and longer scale, it's possible that those populations being extinguished in your protected area by climate change will encounter acceptable conditions scattered about the semi-natural matrix and will succeed at tracking the upslope or poleward movements of their ideal climates, rather than simply expiring at the reserve's edge.

As Crome (1997) states, "If the biota in the fragmented landscape is to persist, then management of the matrix becomes all-important." Other ecologists and conservation professionals increasingly agree with that statement.[6] Many conservation strategies now focus on whole landscapes (Poiani et al. 2000; The Nature Conservancy 2000), including the semi-natural matrix. Returning to figure 1.1, though, how can management decisions be made without the information provided by scientific inquiries?

Box 7.1. Coming Up with Questions about Contents and Context

Consider the protected area where you work or might be interested in working. Either sketch a map or else examine an aerial photo, satellite photo, GIS overlay, or other representation of the protected area *and the landscape surrounding it*. Note the following:

- the shape of the reserve's legal border;

- the extent to which the legal border corresponds to boundaries between unbroken expanses of original habitat and the semi-natural matrix alongside;

- the variety of types of habitat on both sides of the legal border and of the real boundary, especially the different elements of the semi-natural matrix; and

- the sizes and shapes of the patches of different land use.

Now apply the method of box 3.2 to this map and what you've just perceived. That is, frame as many questions as possible about this landscape. Reflect on the questions, then choose those that seem most urgent to real-life conservation concerns and consider designing studies to answer them, following the process presented in chapter 4.

What to Do?

As a conservation professional, if your activities can extend beyond the reserve proper and into parts of the semi-natural matrix, don't dismiss that potentially rich landscape as you follow the "management cycle" of figure 2.4 and propose conservation guidelines. As a field ecologist, consider that questions about phenomena in the semi-natural matrix may be equally intriguing as—and more manageable than—questions about features of the reserve proper. Whether or not your own activities extend into the semi-natural matrix, consider encouraging local communities to articulate their concerns, propose questions, and pursue critical inquiries with yourself as collaborator (see chapter 10). If appropriate, following such inquiries, you might work with the communities to manage their landscape in such a way as to (a) minimize the detrimental edge effects penetrating the reserve, (b) maximize the beneficial edge effects penetrating the semi-natural matrix, and (c) maximize the conservation potential of the semi-natural matrix in itself. For now, at least go through the exercise in box 7.1.

Whether you're working with communities or on your own, does the complexity of a landscape-level inquiry seem overwhelming? Fortunately, some shortcuts exist for inquiring into edge effects or other cryptic effects of human activities on the "ecological integrity" of a landscape. Those shortcuts and others may be useful in broad-scale conservation strategies. Which shortcuts will you choose?

CHAPTER 8

Indicators versus Targets: Shortcuts to the Landscape's "Health"?

The question "What are we monitoring or assessing, and why?" is fundamental to selecting appropriate indicators.

—Reed F. Noss (1990)

Many questions and inquiries related to conservation focus on a particular plant species, animal species, or group of species. Perhaps the primary concern is that species or group itself. If so, you're treating it as a *target species* or *target group*. A biologically (and statistically?) significant pattern in the results of your inquiry signals to you that something's going on *with that target species or group*. During the reflection phase you might also choose to speculate that something might be going on with other species or ecological processes as well, but you won't act further on that speculation: your focus remains on the species you're studying. Chapter 4's question on the effects of selective logging involved three target groups: forest birds, small mammals, and frogs. In chapter 5, Dra. Navideña and Dr. Cumpleaños chose a single target species, the ferocious caiman. You might select a particular target species or group for any one of several reasons, as discussed below.

Perhaps, instead, your conservation concern and your focus are much broader. (1) Your focus might be the entire assemblage of plant and animal species, your concern being the "health" of the assemblage with respect (a) to highly visible phenomena that might affect its robustness, such as proximity to the forest-agriculture boundary, selective logging, road-building, or mining; (b) to more cryptic stresses that you suspect but can't pinpoint; or (c) to the conservation guidelines and strategies you've implemented (e.g., The Nature Conservancy 2000). (2) Your focus might be the "ecological integrity"[1] of the landscape or a particular patch of it, not only the biota but also the full suite of ecological interactions and processes in which the biota participates. As with the species assemblage alone, you may wish to evaluate ecological integrity with respect to known sources of perturbation, less visible stresses, or success of guidelines already implemented. These are large-scale worries. How will you address them? Even a huge, multifaceted, and multidisciplinary conservation project can't possibly address what's going on with every plant, animal, and microbe population in the landscape, let alone every one of the interactions and processes that pertains to ecological integrity. If you're working with only a few colleagues or on your own, directly answering these urgent questions at all scales is clearly impossible.

So, to address questions about the biota's health or the landscape's ecological integrity, you've selected a particular species or species group as a shortcut, a surrogate for many other members of the species assemblage and for the ecological processes that involve them. You assume that responses of what you're calling the *ecological indicator species* or *ecological indicator group* to stresses or strategies will reflect responses of those other species and ecological processes as well (Poiani et al. 2000). A biologically (and statistically?) significant pattern in the indicator species or group signals to you that something's going on *with the biota as a whole, and by extension with the ecological integrity of the landscape* (McGeoch 1998). That is, ideally the indicator species or group tells you either that all's well or all's not well with the biota, or that part of the landscape, or the landscape as a whole.[2] If that's your goal, you must be exceedingly careful to choose the most appropriate indicator species or group, and you must take special pains to select carefully the particular response variable(s) that you'll measure with respect to that ecological indicator.

Ecological indicators are by no means limited to species. Technically, species make up only one component of the landscape's biodiversity (Noss 1990). Inquiries on ecological integrity, including but not restricted to those studies labeled as "biodiversity monitoring," might involve several kinds of biological or physical variables as ecological indicators.[3] Conversely, just because you're quantifying an environmental variable other than indicator species or groups doesn't automatically make that variable an ecological indicator. For example, you might choose to study the concentration of nitrates in lakes that you know to be exposed to different intensities of agricultural runoff as an intriguing phenomenon in its own right—in which case you're treating that variable as a target rather than an ecological indicator. In this chapter I'll focus mainly on species and species groups simply because those are often the least expensive, best understood, and most frequently chosen targets or ecological indicators. I'll also strongly recommend, though, that you consider as ecological indicators—or targets—certain key, easily quantified ecological processes.

At first glance, the difference between viewing species as targets or as indicators may seem trivial. Nevertheless, the two philosophies differ in critically important ways. Furthermore, the species most appealing as targets often make poor indicators, and the best indicators don't often make appealing targets. Regrettably, conservation professionals have tended to slap the label "bioindicator" quite indiscriminately on any species, species group, or other ecological phenomenon that they've chosen to monitor or otherwise investigate, whether it's truly an ecological indicator as defined above or a target instead. At best such subjective labeling may lead to a lot of misinterpretation and a lot of misinformation disseminated in conservation circles, in the popular press, and among financial donors. At worst it can lead to disastrous management decisions. In this chapter I'll suggest some good and bad choices for indicators while emphasizing that choosing target species, though equally valid, involves very different criteria and goals.

Criteria for Choosing a Target Species

Why does a particular species or group of species become the target of management concerns, regional policies, worldwide regulations, or basic ecological investigations? Most possible reasons fit into a simple scheme.[4] One reason a species is chosen as a target is because it's an Important Species—to conservation politics, to conservation priorities, or to economics. It's important to conservation politics:

- as a *charismatic species,* an aesthetically appealing organism that's likely to spark sympathy (and a frenzy of check writing) among the general public—big brown eyes and fur, feathers, fangs, or flippers help a species to win the charismatic label and grace full-color calendars, but some vertebrates with none of those traits (e.g., the red-eyed tree frog, poison-dart frogs), some invertebrates (e.g., certain butterflies), and some plants (orchids in particular) also qualify (question: Might too much emphasis on charismatic species lead the public to believe that landscapes without them aren't worth conserving?);

- as a *flagship species,* a charismatic species that spearheads a conservation effort or management plan for a particular landscape (same question as for charismatic species); or

- as an *umbrella species* with large ecological neighborhoods or an extensive geographic range as a whole, such that conservation measures aimed at the target species will end up conserving a great many other species, and ecological processes, as well (same question as for charismatic species; plus, Will other species' sources and sinks necessarily match those of the umbrella species?).

It's important to conservation priorities:

- as a *vulnerable species* that's currently in danger of going extinct or whose ecological traits suggest that it might soon be in danger—a vulnerable species that wins official recognition becomes a *threatened or endangered species* (TES) (questions: Does a relationship exist between how easily a species can win TES status, or even just the "vulnerable" label, and the local ratio of biologists to species? Is that ratio likely to be highest in landscapes with the most species that truly deserve the label?); or

- as a *keystone species* "whose impact on its community or ecosystem is large, and disproportionately large relative to its abundance" (Power et al. 1996).[5] Some keystone species are easy to spot. Remove elephants from African landscapes they currently occupy, or add them to landscapes they haven't occupied for a while, and you'll see dramatic changes in the landscape's physical and biological features. Remove beavers from North American landscapes, or add them to landscapes in Tierra del Fuego, and you'll see dramatic changes in the landscape's physical and biological features. Others may be much harder to spot at first. Questions: What's a nonkeystone species, and can you easily distinguish it from a keystone species? Might "keystoneness" be a continuous, rather than discrete, trait (Kotliar 2000)? Have all "very keystone" species in any landscape been discovered, properly labeled, and included in conservation guidelines?

It's important, as an *economically valuable species* either to local consumers or in the commercial marketplace. Question: Will species currently labeled as "valuable" and "not so valuable" always retain those same labels, and if not, what will happen to your conservation guidelines?

A second reason a species is chosen as a target is because it can be advertised, although not really used, as an *indicator species*. Almost certainly this is false advertisement (and see McGeoch 1998). Recall the difference in definition, and see the next section. Nevertheless, some people feel that they must justify having selected a target species by extolling, without hard evidence, its properties as an ecological indicator. If you feel that a species is Important to consider for one or more of the above reasons, that's certainly sufficient justification for studying it. If you can't ascribe some "importance" or another to the species, don't twist the truth, employ entirely circular reasoning, and call it an indicator. Be truthful and admit freely that you've chosen it for the third reason a species is chosen: because it's there, it intrigues you, and consequently it's an *interesting species*. Perhaps this label is the most honest of them all. After all, who's to say that some species native to the landscape are more Important than others? Shouldn't conservation involve them all, with some exceptions such as the smallpox virus and *Anopheles* mosquitoes?

Of course, these labels aren't mutually exclusive. A certain target species could be simultaneously charismatic, flagship, umbrella, vulnerable, keystone, economically valuable, and of course interesting. Nor is the judicious use of labels bad, as long as you're truthful about it (Caro and O'Doherty 1999). Still, be very careful when developing conservation guidelines based on the labels (Crome 1997). They may be quite subjective. It's not fair to dismiss those species that haven't (yet) been labeled as vulnerable, or keystone, or economically valuable. It's dangerous to rely on charismatic, or flagship, or umbrella species to distinguish landscapes that are worth conserving from those that aren't (Schwartz 1999; Poiani et al. 2000). Finally, to repeat: don't advertise a target species as an indicator unless you have read the following two sections and have convinced yourself that you're not twisting the truth.

Criteria for Choosing an Indicator Species (or Group)

If your ecological indicator is to live up to its name, it has to be more than just interesting: it must meet certain fundamental criteria. Some criteria for choosing indicators are obvious and relate to points made earlier, especially in chapters 4 and 6. Others, less obvious, relate to the philosophy of the indicator concept and to the quote heading this chapter. As you'll see, few species groups—let alone individual species—can satisfy all the criteria at once. Still, some do better than others.[6] The criteria are as follows:

1. *Objective sampling:* The indicator should be one that you can sample effectively and objectively through direct observations, measurements, or counts with a minimum of sampling bias, using biologically reasonable evaluation units.

2. *Efficient sampling:* The indicator should be one that you can sample efficiently, producing a flow of data during most of the time that you're engaged in the inquiry. That is, it should not require an inordinate amount of setup time before the first useful data emerge.

3. *Sample size:* The indicator should be capable of providing a large number of replicate response units per unit time, effort, or money invested.

4. *Sampling expense:* Sampling the indicator should involve a minimum of expensive apparatus and sophisticated procedures.

5. *Familiarity:* The natural history and taxonomy of the species or group should be well known.

6. *Scale:* Ecological neighborhoods of the indicator species or group, likewise the scale at which any other ecological indicator operates, should correspond to the scale most appropriate to the conservation concern.

7. *Sensitivity:* Your own preliminary data, or trustworthy studies done elsewhere, should already have demonstrated that the indicator is sensitive to factors of conservation concern (e.g., pollution, habitat fragmentation, soil compaction, changes in watershed management).

8. *Aptness as a surrogate:* The indicator species or group should respond consistently to environmental change over time and space either in similar fashion or in directly opposite fashion to much of the remaining biota. In particular, the presence and absence or the population density of a single species, the composition and diversity of an indicator group, should clearly correlate with the landscape's ecological integrity as you've chosen to define that. If the indicator species or group plays a critical ecological role itself, that's an additional bonus—but not an absolute criterion. Alternatively, should the indicator be an ecological process rather than a taxon, the process should involve many species with quite different lifestyles.

9. *Consistency:* The indicator species or group should be equally active or accessible at all seasons when sampling might occur.

10. *General interest:* At least for inquiries that involve the semi-natural matrix, the indicator ideally should respond to factors that also concern rural communities, whether or not the indicator in itself interests them (and see chapter 10).

Single Species as Ecological Indicators?

Sometimes a particular species of animal or plant tends to thrive in landscapes nearly untouched by human activities and to vanish when ecological integrity is disrupted. I can't think of a single case, though, that complies with all the criteria just listed. Like other authors (e.g., Landres, Verner, and Thomas 1988; Noss 1990), I strongly suggest that you *never use one species alone as a positive indicator,* that is, an indicator you expect to be positively related to ecological integrity or to biodiversity by any definition. In particular, most plants and animals native to tropical or subtropical landscapes display geographic ranges that are notoriously patchy regardless of any human influences. While the presence of such a species might truly signal that all's well with the rest of the biota and the landscape's ecological integrity, does its absence signal that all's not well? Not necessarily. Its absence signals that it's absent. That's all.

On the other hand, certain species are "exploiters" (McGeoch 1998) of human-related disturbance. Some opportunistic plants, animals, and microbes follow humans almost everywhere, almost always showing up when "all's not well" with the native biota or the landscape's ecological integrity. Some such species show up even when the disruption is quite cryptic. Thus, they may provide excellent early warnings. These *negative indicators* themselves aren't often dangerous or disruptive. Rather, their presence signals that more subtle, complex, and serious events are occurring with respect to ecological integrity—or even human health. The negative indicator most widely recognized worldwide is the bacterium *Escherichia coli,* whose presence in water bodies signals contamination by human feces. Indeed, *E. coli* complies with most or all of the criteria listed in the last section. Only a few strains of *E. coli* are truly dangerous to humans, but the presence of any strain in some quantity indicates that other, more pathogenic parasites are also likely to be present. Conversely, a very low density or the absence of *E. coli* always indicates low levels of fecal contamination and low health risks—with respect to infection by human gut parasites, at least. Public health authorities often find it easier and cheaper to monitor water quality by sampling for *E. coli* than by other means.

Of course, your conservation concerns probably involve landscape health rather than just human health. What species more visible than *E. coli* to the naked eye might indicate that all's not well by their presence? Although to my knowledge none has been adequately evaluated, candidates might include:

- Bracken fern (*Pteridium aquilinum*). In some landscapes bracken fern lives happily with the rest of the native biota—for example, in the understory of some ancient clones of the aspen trees described in chapter 6. In many moist subtropical or tropical regions worldwide, though, bracken fern thrives only where the landscape's ecological integrity has been disrupted, as through frequent burning or grazing and trampling by cattle. The fern also invades many wet montane forests that have experienced much more cryptic disruptions, penetrating quite some distance along trails and even into treefall gaps.

- Rosa mosqueta (*Rosa eglanteria*). As described in chapter 6, in the national parks of Argentina's lake district this exotic plant invades sites that have been disturbed by grazing, logging, burning, or trampling.

- Other invasive plants. With the possible exception of the wet tropics, almost any landscape in the Western Hemisphere now offers a choice of several exotic, invasive plant species that might function as negative indicators.

- Black rat (*Rattus rattus*). Rats usually associate with human influences that are far from subtle, so why might you need them as indicators? Because sometimes rats also accompany quite cryptic disruptions of the native biota or landscape. If, as occasionally happens, a rat turns up in the live traps you thought were in an "untrodden" landscape, you should worry—and start searching for some form of human intrusion that you hadn't noticed before.

- House wren (*Troglodytes aedon*). The bubbling, cheerful song of the house wren brightens many landscapes, in and out of protected areas, from Canada to Tierra del Fuego. The darker side: in forested regions, hearing the wren's song should warn you that the forest isn't as intact as you might have thought. Some clearing has undoubtedly occurred, probably accompanied by some disruption of the original forest biota and ecological processes.

None of the species I've listed makes an ideal choice. By observing your own landscape, you may be able to identify better candidates. Perhaps, though, you're beginning to realize that there's no such thing as the perfect indicator species—except *E. coli*. Might an ecological indicator composed of many species serve better?

Indicator Groups on Land?

If you're looking for target groups, terrestrial vertebrates often make excellent choices. Vertebrates are often "important" by several criteria, and usually also "interesting." The effects of human activities on bird, mammal, reptile, and amphibian assemblages invoke considerable interest and great concern. Likewise, monitoring of vertebrate assemblages can certainly be a legitimate and important activity in its own right.

Nevertheless, as you've already inferred from the selective logging inquiry, vertebrates rarely make good ecological indicators (Landres, Verner, and Thomas 1988; Hilty and Merenlender 2000). Although frequently the target of conservation initiatives or basic studies, the bird assemblage or the large mammal fauna tends to make an especially poor choice for an ecological indicator group. Small mammals tend to comply with more of the criteria for indicators but still don't signal much about ecological integrity that you wouldn't find out more easily by other means—unless you happen to trap a black rat way out in the "pristine" forest. Medellín, Equihua, and Amin (2000) argue convincingly that bats may constitute an excellent indicator group in the lowland tropics, although their proposal needs to be evaluated in a variety of landscapes. Frogs and sometimes lizards meet more of the requirements for indicators. At present, some conservationists and the popular press tout frogs as "mine canaries," sensitive early warners of worldwide environmental contamination that soon will affect other species including ourselves. The concern should focus on frogs as targets rather than indicators, though. Frogs are definitely in trouble worldwide, but it's not at all clear that the hotly debated causes of amphibian declines will affect other taxa in the same fashion. If your question demands a vertebrate *target* group, often frogs (or lizards) make an excellent choice, as discussed for the selective logging question. If you're seeking a vertebrate *indicator* group on land, though, you may be disappointed.

Bird Guilds as an Alternative to Species Lists

A *guild* is a group of species that use the same kind of resource in similar ways. For example, the nectar-feeding birds in your protected area constitute a guild, and the fruit-eating birds, or fruit-eating birds and mammals combined, constitute another or perhaps two (one eating small fruits and the other consuming large fruits). From a different perspective, birds that nest in holes in trees constitute a guild, while birds that nest in holes in dirt banks constitute another. The avifauna as a whole presents a *guild profile,* an array of different feeding (or nesting) guilds, each with one to many species. Some conservation scientists argue convincingly that even though the avifauna's species composition alone serves poorly as an ecological indicator, its guild profile may be quite sensitive even to subtle effects of human activities.[7] Considering your own landscape, would the guild structure of the avifauna be a valid and useful ecological indicator? How would you determine that?

Butterflies

Like birds and large mammals, butterflies ooze charisma, and like birds and large mammals, they have attracted many larval natural historians who later metamorphosed into some of the world's leading conservation scientists. Butterflies certainly constitute a valid target group. As an ecological indicator group or even as a biodiversity indicator, though, they haven't met with resounding success.[8] Butterflies' responses to habitat variables and effects of human activities are extraordinarily complex, partly because it's difficult to distinguish individuals "just visiting" from individuals that truly "belong" and partly because butterfly larvae and adults have vastly different lifestyles and resource requirements. Butterflies often comply with the first five criteria for ecological indicators but rarely with the last five. The lesson here: Just because conservation scientists have a special fondness for a particular animal (or plant) group doesn't automatically qualify that group for indicator status.

Dung (and Carrion) Beetles

The habits of these insects, belonging to certain subfamilies of the beetle family Scarabeidae, aren't especially charismatic (figure 8.1), but the beetles themselves can be amazingly so. Furthermore, evidence is accumulating that dung beetles closely reflect the ecological integrity of the above-ground habitat in many tropical and subtropical landscapes.[9] They're rapidly graduating from target to indicator status, and as they gain in popularity as well as usefulness you're likely to have expert help nearby (see appendix C). You can sample dung beetles objectively, efficiently, and cheaply with baited pitfall traps. All you need is a good supply of plastic soft-drink cups and fresh fecal material. Dung beetles even comply well with the criterion of general interest. As almost any farmer recognizes, without dung beetles much of the semi-natural matrix of the Western Hemisphere would now be ankle deep in cow manure.

Other Beetles

Beetles occur nearly from one pole to the other. Some landscapes support many thousands of species. Dung beetles may come closest to being the ideal indicator group, but some other beetle groups may also comply with many or most of the criteria listed above. Assemblages of tiger beetles (family Cicindelidae) may indicate ecological integrity in many tropical and subtropical landscapes of Latin America and elsewhere (Pearson and Cassola 1992; Rodríguez, Peason, and Barrera 1998). Other, smaller beetles that live in leaf litter may also make good indicators in some forest habitats (see Nilsson et al. 1995; Rykken, Cupen, and Mahabir 1997). Finally, the darkling beetles (family Tenebrionidae) might serve as excellent ecological indicators in arid and semi-arid landscapes (personal observation), although to my knowledge they have not been thoroughly evaluated in this context.

Other Insects and Relatives

Figure 8.1.
A classic display of dung beetle behavior
(Monteverde, Costa Rica).

Other leggy invertebrates might also make good indicators.[10] Your choice should depend on how well you and your contacts know their natural history. Don't assume that invertebrate groups will automatically make good indicators, though. Ants, whose biomass and diversity in many tropical or subtropical landscapes put vertebrates to shame, don't always comply with the criteria for indicator groups (Andersen 1997).

Invasive Plants

Almost every landscape has a diverse suite of invasive plants that includes not only the weedy exotics noted earlier but also a number of opportunistic natives quick to exploit disruptions in the vegetation cover. Regardless of whether any one such plant serves as a negative indicator species, the entire group of plant invaders might comply with all the criteria listed and make an excellent "negative indicator group." In this case, increased abundance or diversity of plants in this group would signal diminished ecological integrity.

Earthworms (or Nematodes)

Are you interested in an animal group that both reflects and contributes to ecological integrity below ground? If you work in a region of moist soils, grab two or three gunnysacks, sling a shovel over your shoulder, and go dig worms (figure 8.2). You'll be following the lead of local farmers, who traditionally relate earthworms to soil quality. In many landscapes, earthworms may comply superbly with all the criteria for indicators except the criterion of familiarity.[11] Most humid landscapes present a suite of native earthworm species and several exotics, the exotics usually signaling some disruption of the ecological integrity of the soil environment. Furthermore, although it's not easy for many of us to tell one earthworm species from another or to distinguish exotics from natives (criterion 5), expert help may be nearby (appendix C).

In any event, you don't necessarily need to identify earthworms in order to use them as an indicator. The total numbers and biomass of earthworms per evaluation unit, such as a chunk of rapidly excavated soil $40 \times 40 \times 20$ cm in volume, may provide decent response variables. Because size varies

Figure 8.2.
Sampling earthworms (Coroico, La Paz Department, Bolivia).

a lot among earthworms, these two measures may present quite different patterns. The only serious drawback to earthworms is that you may encounter extraordinarily high variability among evaluation units, so you might need to subsample extensively within each response unit. On the other hand, earthworms offer one great advantage over almost any other indicator group: multiple use. If there's a nearby river or lake, at the end of the day's work you're all set to go fishing.

Recently, Bongers and Ferris (1999) proposed soil nematodes as the epitome of an ecological indicator group. As the authors point out, it isn't too difficult to find nematodes, as four out of every five multicellular organisms in the world belong to this group. It's quite possible that nematodes in any landscape comply with all the criteria listed above except for criterion 5 (familiarity) and perhaps criterion 4 (low expenses). If you have access to experts at an agricultural experiment station, though, nematodes may comply with those criteria as well. The only other drawback may be that they're too small to use as fish bait.

Indicator Groups in Water

Speaking of fishing, do your conservation concerns include rivers, streams, ponds, and lakes? If not, they should. The ecological integrity of water bodies and whole watersheds is at least as important to conservation as is the ecological integrity of your landscape's terrestrial habitats, and the two are intimately related (Allan and Flecker 1993). Often, water quality is monitored with chemical and physical indicators. Such variables may not reflect ecological integrity in the broad sense or biological integrity in specific (Karr 1991, 1992; Karr and Chu 1998). Also, chemical and physical techniques are often expensive. Furthermore, they reflect the status of the water at the moment it's sampled, rather than its recent history. A devious polluter could release chemical contaminants into a fast-flowing stream at midnight, and by the time you sampled the next morning, few traces might remain. Fortunately, though, the water bodies in your protected area almost certainly provide several candidates for ecological indicator groups, from bacteria or diatoms to vertebrates. Two taxa in particular stand out: insects and fishes.

Benthic Insects

Sample the bottom of an uncontaminated, rocky-bottom stream anywhere in the world and you'll find a number of different insects (figure 8.3). These often include nymphs of mayflies (order Ephemeroptera) and stoneflies (order Plecoptera), larvae of caddisflies (order Trichoptera) and many kinds of "true" flies (order Diptera), and both larval and adult beetles (order Coleoptera), as well as many other insects. Slower-moving rivers, and ponds and lakes, may contain different species of those same orders plus others, such as nymphs of dragonflies and damselflies (order Odonata). These benthic or bottom-dwelling insects vary widely in life cycle, microhabitat preference, mode of locomotion, food preference, and feeding mode. They also vary extraordinarily widely in sensitivity to changes in water quality.[12] For example, many can't tolerate even the slightest compromise in oxygen content or chemical purity of water, while some fly larvae thrive in extraordinarily contaminated, oxygen-deficient water.

The sampling methodology for benthic insects varies with the type of water body. For example, to sample shallow, fast-flowing streams with rocky bottoms, you can either construct a "kick net" from nylon mosquito netting and two sticks or you can purchase its more precise cousin, the Surber stream bottom sampler. Aside from a basic stereomicroscope (dissecting microscope), you can obtain

Figure 8.3.
Benthic insects sampled from a stream near San Carlos de
Bariloche (Río Negro Province), Argentina.

all the other supplies cheaply and easily. Most references I've listed and numerous others will guide you through the simple sampling procedure. In a short time local experts (see appendix C) can train you to distinguish the different insect orders and to distinguish "morphospecies" from one another.

If sampled correctly in a well-designed study, benthic insects comply well with all criteria for a bioindicator group. They offer at least two unique advantages as well: (1) sampling benthic insects may cost much less than sampling the chemical or physical properties of water (Karr 1998); and (2) the assemblage of benthic insects you find at a site reflects not only today's water quality but also the cumulative effects of yesterday's, last week's, and to some extent last month's or, in extreme cases, last year's. Given the complexity of taxa, lifestyles, and sensitivities among benthic insects, you might select any one of a great number of response variables to measure. Which variables are most suitable for indicator purposes?

The Benthic Index of Biological Integrity

In North America, James R. Karr and colleagues have developed a painstaking, objective methodology for deriving a single response variable, the Benthic Index of Biological Integrity (B-IBI), to be used at different sites within a given watershed. The Index of Biological Integrity is actually a composite measure that incorporates a number of different ecological or taxonomic variables, each recorded directly from the samples of stream insects.[13] For example, the B-IBI for a particular watershed might combine values for the species diversity of various sensitive taxonomic groups, and values for the diversity or numerical dominance of certain ecological groups such as predators or filter-feeders.

The Ephemeroptera-Plecoptera-Trichoptera Index

You might not have the time, resources, or experience to go through the lengthy process of developing the Benthic Index of Biological Integrity. Nevertheless, one of the directly measured variables usually incorporated in a B-IBI often makes a valid and reliable indicator by itself (see Wallace, Grubaugh, and Whiles 1996). This is the Ephemeroptera-Plecoptera-Trichoptera (EPT) index. Simply tally the total number of morphospecies in your evaluation unit that belong to the three easily recognized insect orders Ephemeroptera (mayflies), Plecoptera (stoneflies), and Trichoptera (caddisflies). Many species in these three groups are highly sensitive to pollution or other disruptions of ecological integrity, others are slightly less sensitive, and still others can tolerate moderate levels of contamination. Consequently, the total number of morphospecies you record in a sample declines as ecological integrity deteriorates. My colleagues and I have found the EPT index to be a useful and rapidly assessed response variable from Colombia southward.

Please heed two warnings about this response variable. (1) The diversity of stoneflies declines near the equator. In northern South America, even the healthiest running waters often hold a single stonefly species at most. There, the EPT index essentially becomes an ET + 1 index. (2) Especially in northern Andean streams where trout have been introduced, insect densities per standard-area evaluation unit (for example, one Surber sample or one kick-net sample) may be orders of magnitude lower than in many streams in North America or the southern Andes. I suggest that you define your evaluation unit as a certain minimum number of benthic insects no matter how much effort or stream-bottom area is involved in getting those. In my own experience, ≥ 300 insects of all orders, or ≥ 500 individuals in streams with numerous larvae of true flies (Diptera), often serve as an acceptable evaluation unit.

The Index of Biological Integrity for Fishes

In many Central and South American watersheds, the fish assemblage consists of one to three species of introduced trout and an occasional survivor from the preexisting native fish species.[14] Lowland river systems, though, may support very diverse fish faunas. Could the fish fauna serve as an indicator of ecological integrity? Decades ago, Karr and colleagues successfully developed the IBI concept for fish assemblages in North America (see appendix C) and then used values of the fish IBI to indicate ecological integrity of rivers and river systems. The fish IBI also shows promise for watersheds in Mexico (Lyons et al. 1995) and Colombia (S. Usma, personal communication). If the river systems in your landscape include a diverse fish assemblage and if you have contacts with ichthyologists or fisheries experts, these aquatic vertebrates might serve as an excellent indicator group.

Back on Dry Land: A Terrestrial Index of Biological Integrity

As this book is being written, in North America the concept of the Index of Biological Integrity is being applied to terrestrial arthropods. Conservation scientists working with Karr (see appendix C) and others are currently formulating ways to produce sensitive indices to ecological integrity by integrating many different response variables, spanning a range of taxonomic or ecological groups of insects and other invertebrates.

Ecological Interactions as Indicators

Your landscape's ecological integrity involves more than just a species list. It also involves what those species do to, or with, each other. Consider these three basic classes of ecological interactions:

- *exploitation* (including predation, parasitism, herbivory, and seed predation), in which some organisms use the energy and nutrients in others to the detriment of the latter;

- *mutualism,* in which some organisms use others but are used by those in turn, sometimes leading to a mutual benefit;

- *decomposition,* in which some organisms use the energy and nutrients in the leavings or corpses of others.

Such processes may have just what you're looking for in an ecological indicator. Each is an essential component of ecological integrity. For example, if patterns in seed predation, pollination of flowers, or leaf litter decomposition veer away from those values typical of undisturbed landscapes, that signals that numerous other ecological interactions and processes must be veering away as well. In addition, each process mentioned indicates the health of a substantial part of the biota. If patterns in seed predation diverge from "normal" across plant species, that signals a change in the nature of the large array of insects and the smaller set of vertebrates that consume seeds. If patterns in flower pollination diverge from normal across plant species, that signals a change in the composition and behavior of an even vaster array of insects and, in many tropical habitats, many vertebrates as well. If the rate of leaf litter decomposition slows down, that signals a change in the composition and activity of a still vaster array of bacteria, fungi, and invertebrate animals involved in the decomposition process. Finally, changes in the nature of these ecological processes may have serious consequences for the fate of the habitat patch or the entire landscape. For example, changes in rates of seed predation or pollination may affect the nature of the next generation of plants and over time alter plant (and animal) species composition. Changes in decomposition rates will affect soil nutrients and the physical environment at the soil surface, altering the patterns of survival and germination among seeds and again affecting the fate of the entire habitat. Let's examine each interaction more closely.

Exploitation

Animals prey on or parasitize other animals, a few plants parasitize other plants, and a great number of animals exploit plants or plant parts as food. The last set of interactions in particular provides a number of possible ecological indicators.

Predation on Seedlings. Many plant seedlings end up as snacks for vertebrates and sometimes invertebrates. Rates of seedling predation may vary with factors of conservation concern (Dirzo and Miranda 1991).

Herbivory on Leaves. Nearly all vascular plants lose parts of leaves or whole leaves to vegetarian insects (figure 8.4), other invertebrates such as slugs, and vertebrates—not to mention loss to bacterial, viral, and fungal pathogens. Rates and kinds of leaf damage may vary with factors of conservation concern.

Seed Predation, Predispersal. Seeds of virtually all plants in your landscape—natives, exotics, and

Figure 8.4.
Herbivore damage to leaves of a
Cecropia obtusifolia sapling (Monteverde,
Costa Rica).

crops—risk predation at two different stages: when still on the mother plant, and after landing on the soil. Larvae of one insect group, the seed beetles (family Bruchidae), prey on seeds of a great number of tropical and subtropical plants (figure 8.5) before and after seed dispersal but especially before. Many other insects also participate in predispersal seed predation. The frequency of seeds attacked before dispersal could serve as an excellent ecological indicator in any landscape but especially in a warm, semi-arid one, where pods of the omnipresent legume plants and fruits of many other plants openly display how many seeds they've lost.

Seed Predation, Postdispersal. Even if seeds survive until they land on the soil, they're still far from safe. Fungi, bacteria, insects, rodents, and birds consume many, and the rate of consumption may

Figure 8.5.
Predispersal seed predation. Many of these fruits, collected
from plants in the Chaco Serrano of Argentina (Tucumán
Province), have tiny holes on the outside. Each hole indicates
the presence of a beetle larva munching on the seed inside.
Photograph by Marcelo A. Aizen.

Figure 8.6.
Postdispersal seed predation assessed by following the fate of
piles of one hundred beans set out on the forest floor
(Imbabura Province, Ecuador).

vary with factors of conservation concern. You could assess this ecological indicator either with seeds of native species or by an experimental approach (figure 8.6).

Mutualism

Of course, not all animals end up harming the plants whose products they consume. Sexual reproduction of most plants in the tropics and subtropics depends on pollination by animals that visit flowers for nectar, pollen, or other products. The fleshy fruits that many plants produce are often consumed by animals—including some very charismatic vertebrates—that often discard or defecate the seeds away from the mother plant, thereby effecting seed dispersal.

Plant Reproduction. The pollination rate of individual flowers examined within plants, among plants, and across plant species can make an excellent indicator variable, with one qualifier.[15] You need an expensive microscope, some other laboratory equipment, and quite a bit of time. Instead of examining pollination alone, though, you might assess a different, equally important response variable: the frequency of fruits or seeds matured per flower initiated (figure 8.7). The rate of fruit and seed production per flower reflects not only the events at the time of pollination but also all other events that might take place between fertilization and fruit maturation, such as fruit abortion by the plant itself (often induced by physical stress or nutrient limitation) and parasitism or predation of the immature fruit. All you need to do is to mark a large number of flowers and then return during the season of fruit maturation.

Fruit Consumption and Seed Dispersal. It's nearly impossible to follow an individual seed once it leaves the maternal plant in the mouth or gut of a fruit-eating animal. You can take an indirect approach, though, by simply recording the rate at which ripe, fleshy fruits disappear from their

Figure 8.7.
Evaluating one stage of plant reproductive success: flowers marked to assess the proportion that produces mature fruits (Monteverde, Costa Rica).

mother plants. Many ecologists assume that on average a fruit's disappearance signifies that it was consumed by those vertebrates that are likely to discard or defecate the seeds alive. Therefore, this extraordinarily straightforward variable may serve as an excellent indicator of ecological integrity, in particular the status of the suite of vertebrate fruit-eaters and the long-term persistence of the plant populations involved. All you need to do is to mark a certain number of nearly mature fruits and then return a set number of days later to tally the proportion of those that have disappeared. Wright et al. (2000) followed the fate of palm fruits that had already landed on the soil under the mother palm. The fruits experienced lower removal rates at sites with frequent poaching of large mammals than at sites without poaching. You could consider reversing the logic—after exhaustive preliminary studies—and use rates of fruit removal as indicators of ecological integrity overall and the stresses experienced by the mammal fauna in particular.

Decomposition

From bacteria and fungi through flies, beetles, and vultures, many of your landscape's species are involved in the crucial process of decomposition. Their activity may vary with physical conditions—for example, temperature and humidity—that local human activities could influence. Some methods to measure decomposition rates directly are sophisticated and costly. Fortunately, some very inexpensive, objective, and useful methods also exist.

Decomposition of Fecal Matter. Most landscapes have first-rate providers of experimental material: cows. You could, for example, distribute fresh cow patties across the landscape in accordance with your study design—or have the cows distribute them for you—and record the rate at which each patty gradually disappears. Dung beetles may be responsible for much but not all of the result.

Decomposition of Carrion. If you have access to waste from a slaughterhouse or to a good source of, say, dead rats, then you can examine this even less charismatic yet equally critical indicator process. If you wish to include the role of vertebrate scavengers in the interaction you're assessing, leave the carrion uncovered; if you wish to assess decomposition only by smaller beasts such as beetles, flies, and bacteria, then keep the carrion inside well-secured cages of wire mesh.

Decomposition of Leaf Litter. As long as there's a balance accessible, you can examine leaf litter decomposition through packets made up of fallen leaves packed into mesh bags such as those

Figure 8.8.
Evaluating the rate of decomposition by means of mesh bags filled with leaf litter (Imbabura Province, Ecuador).

found in your local vegetable market (figure 8.8). Dry and weigh each packet before setting it on the ground, and again after a standard time interval of several months. Similar techniques exist for following leaf litter decomposition in streams or lakes.

Out of the Forest?

Some of the ecological interactions described above may also interest the people inhabiting the semi-natural matrix (see chapter 10). For example, consider working with local communities to design an inquiry into seedling predation, leaf herbivory, fruit or seed production per flower, predispersal seed predation, and postdispersal (postplanting) seed predation, all with respect to some design factor such as distance to forest edge—but in crop plants on the converted side of the border instead of understory plants on the forest side. Consider examining rates of decomposition (fecal matter, carrion, leaf litter) with respect to various classes of land use in the semi-natural matrix, use of pesticides, and other practical design factors.

Once More, Targets or Indicators?

Reflect one more time on the selective logging question of chapter 4. Are you truly interested in the effects of selective logging on forest birds, small mammals, and frogs only? If so, you've chosen the targets and designed the study around them, paying heed to the warnings in chapters 4–6. If you're really more interested in the effects of selective logging *on the forest's ecological integrity as a whole,* though, you must revise the study design yet again. None of the three vertebrate classes, even frogs, makes as adequate an indicator group as, say, earthworms, dung beetles, some other beetle group, or perhaps nematodes. Any of the ecological interactions listed above, including those in which mammals or birds participate, probably makes a much better ecological indicator than would the vertebrates themselves.

Deciding whether your question involves targets or indicators is a crucial choice. If you vote for indicators, deciding which indicators to use is the next crucial choice. How will you make that decision? To date, choosing ecological indicators has sometimes been a haphazard process. Few if any workers have explained clearly why they chose one ecological indicator instead of another, other than their personal experience with the group or the process chosen. Recently, McGeoch (1998) and

Box 8.1. Practice with Choosing Indicators

Review the questions you proposed for box 7.1, or if appropriate those for boxes 4.2 and 6.1. Choose a question that addresses concerns about the landscape's ecological integrity or simply the "health" of the native biota as a whole. If you hadn't specifically proposed such a question, rephrase one so that it addresses one of these concepts. Now list the five most useful indicator processes, groups, or species that you might use to address the question. Are they complementary to one another? If not, think again. Finally, for each choice list and describe completely the particular response variable(s) that you would measure. Refer to chapter 4 as necessary.

Hilty and Merenlender (2000) have developed objective, comprehensive criteria for selecting indicator groups, if not yet indicator processes. Let's hope that conservation professionals follow these criteria in the future.

For your own landscape, if possible, choose several ecological indicators instead of just one (box 8.1). If you're restricted to working with only one indicator, make sure it's either a species group or a process, not a single species. Choosing the indicator group or process is only the first step, though. Just as in the selective logging example, you must pass through the entire design process. What's your design factor, and what levels will you examine? How and where will you find the response units that represent the control or baseline situation of 100 percent ecological integrity? What will you choose for response variables? Will the response variable that you analyze be a composite or synthetic statistic like the Index of Biological Integrity? If so, how will you decide on the most appropriate way to select and combine the directly measured variables that contribute to the composite statistic? If you choose a single indicator group, such as dung beetles, what will you record? The total abundance, the biomass, the proportion of native species, the number of individuals of exotic species? Most likely you'll choose a response variable having to do with "species diversity."

CHAPTER 9

Species Diversity: Easy to Quantify, but What Does It Mean?

Species diversity has become a nonconcept.

—Stuart H. Hurlbert (1971)

Beginning with chapter 4, where we were concerned about the possible effects of selective logging on forest vertebrates, and continuing through chapter 8, I've often suggested using species diversity as the response variable—without saying how you might measure it. In fact, most questions that involve possible changes in a target group of species, or questions about the landscape's ecological integrity as reflected by an indicator group of species, employ some measure of species diversity. In theory, biological diversity (biodiversity), as defined broadly, includes much more than just the number of species your landscape supports (Noss 1990). In practice, though, many conservation professionals continue to think of biodiversity in terms of species diversity. Because the idea of biodiversity or species diversity infuses so many conservation concerns and ecological field studies in so many ways, the concept merits its own chapter. This chapter includes some history, so that you'll appreciate the unlikely origin of some common measures and perhaps question the wisdom of their unchallenged use in conservation.

One fundamental assumption underlies the use of species diversity measures, whether to represent biodiversity in the broad sense, to evaluate ecological integrity, or to quantify the diversity in target groups: *species diversity decreases when ecological integrity is compromised.* Later on, you'll have reason to question the validity of this assumption. For now, let's begin by discussing different ways of measuring diversity.[1]

Expressing Diversity as Species Richness

Let's say you've sampled herbaceous plants in three different patches of vegetation (table 9.1), each about ten hectares in area. In each patch you sampled one belt of standardized dimensions, 5 m × 100 m. At Site 2 you encountered more bare ground and fewer plants than at the other two sites. You now wish to characterize the species diversity of each site with a sample statistic, as defined in chapter 4. Should you simply count the number of species and report diversity as *species richness* (*S*)? Perhaps. Had your census belt included all or nearly all of a patch's area, you could be reasonably confident that you'd tallied all or nearly all of the species in that patch,[2] and *S* might be the most straightforward descriptor of diversity. Alternatively, you might choose *S* even knowing it to be an underestimate, if you were confident that the extent of underestimation was consistent among the three patches.

Do the data in table 9.1 truly inspire confidence that Site 1 is more diverse than Site 2? The total

Table 9.1. Some alternatives for quantifying species diversity. Numbers of individuals per species (n_i) sampled in each of three sites, with p_i in parentheses, where $p_i = n_i/N$. For each site, species diversity can be quantified by the total number of species S or by any one of many indices, such as the Shannon-Weaver index H', whose value depends on the log base used, or the inverse Simpson index C_{inv}, whose value does not. See text of chapter 9 for equations. Species A–K are natives (n), while species L–R are exotics (e).

Species	Site 1	Site 2	Site 3
A (n)	160 (0.409)	30 (0.154)	10 (0.028)
B (n)	80 (0.205)	40 (0.205)	0
C (n)	60 (0.153)	80 (0.410)	20 (0.056)
D (n)	50 (0.128)	10 (0.051)	6 (0.017)
E (n)	20 (0.051)	25 (0.128)	0
F (n)	10 (0.026)	3 (0.015)	6 (0.017)
G (n)	6 (0.015)	5 (0.026)	20 (0.056)
H (n)	1 (0.003)	0	0
I (n)	1 (0.003)	0	0
J (n)	1 (0.003)	1 (0.005)	0
K (n)	1 (0.003)	0	4 (0.011)
L (e)	0	1 (0.005)	60 (0.169)
M (e)	0	0	60 (0.169)
N (e)	1 (0.003)	0	50 (0.140)
O (e)	0	0	40 (0.112)
P (e)	0	0	30 (0.084)
Q (e)	0	0	30 (0.084)
R (e)	0	0	20 (0.056)
Total individuals (N)	391	195	356
Total species (S)	12	9	13
C_{inv}	3.96	3.94	8.75
$H'(\log_e)$	1.64	1.60	2.31
$H'(\log_{10})$	0.711	0.697	1.00

number of individuals you encountered, N, in the census belt was twice as great for Site 1 as for Site 2. Both samples included exactly the same set of species whose numbers of individuals, or n_i, exceeded 1. At Site 1, though, you also encountered five species with $n_i = 1$, while at Site 2 you encountered only two such species—which accounts for the difference in S. Are you sure that those three additional rare species are truly absent from Site 2? Might they be absent from your sample simply because you didn't encounter very many individuals in total? This doubt exemplifies a fundamental concern in sampling from species assemblages: unless you census every square or cubic centimeter of the site or landscape of concern, your sample will almost certainly miss some rare species. The smaller your sample with respect to that complete census, the greater the number of rare species you'll likely miss by chance alone. If a given rare species shows up even once in your sample, you know it's present at the site; but if it fails to show up in your sample, you can't conclude that it's absent from your site. *Therefore, you cannot ascribe any special biological meaning to either the number or the identity of those rare species that do occur in your sample.* That is, the difference between 0 captures and 1 capture is often just chance (also see Colwell and Coddington 1994).

Quantifying Diversity with Indices

Perhaps, then, you should choose some measure of species diversity that isn't influenced so much by rare species whose presence or absence on your list might just be an artifact of sampling. You might also prefer a statistic other than S for a different reason. For decades (see below) ecologists have felt that a species diversity measure should have a higher value when all species are about equally common than when great disparities exist in their abundances. If most individuals in a sample of S species belong to a single species, while the remaining species have only two or three individuals each, by this definition the sample is less diverse than one also having S species but with lower numerical dominance, that is, greater *equitability,* among species. Here, for example, S scarcely differs between Sites 1 and 3, but 40 percent of the individuals at Site 1 belong to a single species, whereas individuals are distributed much more equitably among Site 3's species. Do you feel that Site 3 is the more diverse of the two? Most field ecologists would say so, if only because their professors, their textbooks, and several decades of technical papers have told them so. Now, wouldn't it be nice if you could condense all these considerations into a single statistic whose value would increase as either S increased or as equitability increased, a *species diversity index?*

In the 1940s and 1950s, ecologists casting about for just such a magic number stumbled upon some measures invented for use in a very different field, information theory. Those statistics or indices, and others that mathematical ecologists themselves began to invent, seemed ideal for representing species diversity, especially for use in the developing areas of systems ecology, theoretical ecology, and ecosystem energetics. For example, species diversity as measured by some indices was thought to represent the diversity of alternative pathways by which energy might flow through an ecosystem, which in turn, according to the theory and mathematical models then in vogue, could be related to ecosystem stability.

About that time many ecologists less oriented toward theory and more oriented toward natural history began to wonder about species diversity in other contexts—for example, why some sites in the same landscape, or some landscapes at different latitudes, had more species than others. The "information-theoretic" indices and others appeared to be useful ways to quantify species diversity for use as a response variable. Soon the indices were employed widely in field ecology. Shortly afterward, because of new legislation, the late 1960s and early 1970s saw an explosion of interest in assessing environmental impacts of all kinds. Environmental policy makers, environmental engineers, applied ecologists, and environmental consulting firms borrowed species diversity indices from basic ecology and applied the statistics to ecological indicator groups as a convenient way to represent ecological integrity and measure the severity of environmental impacts. Rumor has it that some management guidelines in the United States went so far as to specify the numerical values of species diversity indices, applied to benthic insects or other invertebrates, that marked the cutoff between "all's well" and "all's not well" with the health of water bodies. Other guidelines specified that industries or management protocols should act in such a way as to "maintain or increase H'."

You may already be acquainted with H', the first of two species diversity indices I'll discuss. Both indices are based on the proportion of an assemblage's, or sample's, individuals that belong to each species i, or

$$p_i = n_i/n \qquad (9.1)$$

where by definition $\sum_{i=1}^{S} p_i = 1.0$. The Shannon-Weaver index H' is calculated as

$$H' = -\sum_{i=1}^{S} p_i \log p_i \qquad (9.2)$$

In theory, the logarithms could use any numerical base. For practical reasons, mathematical reasons, or no clear reason at all, different workers have chosen to calculate logarithms to the base 10, the base 2, or the base e (natural logarithms). Values of H' range from 0 (for a sample with only one species) on up, in a rather nonintuitive way. Table 9.1 reports values for H' using $\log_{10}$ and $\log_e$. Note that the values using $\log_e$ are higher. Also note that these indices accomplish what you sought: the values reflect the distribution of individuals among common and moderately rare species rather than the influence of those very rare species that are or aren't in the sample by chance alone. Despite the difference in S, Site 1 and Site 2 have nearly identical values of H', whereas Site 3, with greater equitability among the abundances of the different species, has a considerably higher value.

It's easy to calculate H' on a hand calculator, and even easier to calculate the second index, the "inverse Simpson index" or C_{inv} (also known as the reciprocal Simpson index or Hill's N_2 index). Technically, the exact equation is

$$C_{inv} = \cfrac{1}{\sum_{i=1}^{S} \left[\dfrac{n_i(n_i - 1)}{N(N-1)} \right]} \qquad (9.3)$$

A much simpler approximation is

$$C_{inv} = \cfrac{1}{\sum_{i=1}^{S} (p_i^2)} \qquad (9.4)$$

With a reasonable sample size, say $N > 50$, the difference between using equation 9.3 and equation 9.4 is trivial, so equation 9.4, being the simpler, is the method of choice. C_{inv} varies from 1.0, for a sample with only one species, to S when all S species have exactly the same number of individuals. Return to table 9.1. Again, by C_{inv} Sites 1 and 2 have about the same diversity, whereas Site 3 is by far the most diverse.

These two indices, H' and C_{inv}, don't respond in identical fashion to changes in S or in equitability. As the values in table 9.1 suggest, C_{inv} is a bit more sensitive to changes in equitability, whereas H' is somewhat more sensitive to changes in S. Because of tradition and because of some clever mathematical tricks based on it, H' continues to be by far the more widely used of the two. For almost three decades, though, many ecologists and mathematicians have pointed out that C_{inv} is the better choice. Practically speaking, C_{inv} is not only quicker to calculate than H' but it's also easier to interpret. For example, the values for C_{inv} in table 9.1 tell us that the distribution of individuals among species is about as diverse as if there were about four equally common species in Sites 1 and 2 or nearly nine equally common species in Site 3.

Of course, species diversity indices aren't limited to the two presented here. Beginning in the 1950s, ecologists and mathematicians proposed numerous additional indices, each with slightly different properties or purposes. In what should have been a classic paper, Hurlbert (1971) questioned this practice and proposed some eminently sensible measures based on biology rather than on information theory, systems theory, or mathematical cleverness. Hurlbert was a voice crying in the wilderness, though. Most other ecologists were content to quote his memorable comments (as I've done above) while continuing to use H' and other familiar indices as before.

Misrepresenting Diversity with Indices

It's understandable that conservation professionals, basic ecologists, and ecological consultants have yearned for a tidy statistic to represent species diversity. These indices and others seemed to be the answer, so we've tended to downplay or overlook their drawbacks. The drawbacks are serious, though, when indices become involved in conservation issues. *Using species diversity indices alone to evaluate indicator groups and ecological integrity could lead to grave errors in management decisions.*

An Index Is Just a Sample Statistic

Almost always you're working with data collected from an evaluation unit that's smaller than the response unit whose species diversity you wish to characterize. The value you calculate, whether H', C_{inv}, or some other measure, is an estimate of the true value for that response unit, just as $\bar{x}$ is an estimate of μ. The estimate may or may not be accurate. Magurran (1988) and other sources discuss techniques for calculating the standard error, and subsequently the confidence interval, of most species diversity indices. Nevertheless, in practice many field workers ignore these details, concluding, for example, that the difference between an observed H' of 0.84 for one sample and 0.79 for another sample has biological significance. Unfortunately, this is especially true in conservation-related studies. To see how wrong such a conclusion could be, go to a nearby store, buy some candies, and go through the exercise in box 9.1.

H' Might Misconstrue Species Diversity

Again, as table 9.1 shows, the value of H' for a given sample depends on which base of logarithms you'd chosen for the calculations (equation 9.2). In the early years of H', many researchers reported values for this statistic without reporting the base of logarithms they'd used. Unless the raw data were presented as well, readers had no way of comparing values from one study to another. For example, let's say one worker samples Site 1 in table 9.1 and reports an H' of 0.711, but a second worker examines Site 2 and reports an H' of 1.60. Would your immediate thought be that Site 2 is

Buy a mixture of several hundred small wrapped candies, all the same brand (the same shape and same texture of wrapper) but including six different flavors or wrapper colors. Count out the following "species assemblage" and pour it into a bowl:

100 candies of flavor (color) A

70 candies of flavor (color) B

30 candies of flavor (color) C

5 candies of flavor (color) D

1 candy of flavor (color) E

1 candy of flavor (color) F

Mix the candies thoroughly. The bowl now represents the response unit whose species diversity you wish to characterize. To do so, you'll sample from it. Unbeknownst to you, the true species diversity for the response unit as a whole is $S = 6$, $C_{inv} = 2.71$, and H' (using $\log_{10}$ values) = 0.495.

Now close your eyes and draw at random just two candies from the bowl. What colors are they? What's the S of this sample? Is this a good estimate of the true S? If you draw only two, or three, or five candies from the bowl, what's the maximum value that S could possibly take in each case? Even if every candy of the five were of a different flavor, which is very unlikely, would you accurately estimate S? Return the candies to the bowl. Close your eyes again and draw six candies. Did you attain the real S of 6? Almost certainly you didn't. What does this tell you about the validity of measures of species richness based on small samples?

Close your eyes, stir all 207 candies again, and draw, but this time draw a sample of 20. Open your eyes. Calculate S for your sample, C_{inv} (equation 9.3 or the crude approximation of equation 9.4), and H' using $\log_{10}$ values (equation 9.2). Record these statistics. Return all candies to the bowl. Repeat the blind sample of 20 candies at least three times, always returning the candies to the bowl to maintain the species assemblage of 207 as specified above. Do the values for your sample statistics (indices) always, or ever, match precisely the true values for the response unit as a whole? Do they match each other? Is Candy A, numerically dominant in the response unit as a whole, numerically dominant in every sample? What if you had performed a real inquiry, sampled from three sites you suspected to have different diversities, and obtained three values as different as those you've just obtained for three samples drawn from a single species assemblage? Would you have attached some biological, or conservation, significance to those differences? Would that be wrong, and if so, what might be the consequence of basing management decisions on such results?

If you have time, repeat the last procedure but with samples of 40 instead of 20. Do the new sample statistics estimate the true values better than the old ones? Are the indices based on samples of 40 less variable among themselves than those based on samples of 20? Are you convinced yet that species diversity indices are just sample statistics, and that they're not likely to reflect accurately the underlying species assemblage, simply because of the vagaries of random sampling? It gets worse. In this exercise any one candy, regardless of its flavor, has the same probability of being drawn as any other. Would the same hold true if candies of certain flavors were smaller than candies of other flavors, or more slippery, or heavier such that they tended to sink to the bottom of the bowl? Might you "capture" those candies less easily than the larger, less slippery candies toward the top of the mixture? In this case might your sample and the indices you calculate be not only variable but also biased? You might try answering these questions by purchasing another brand or two of candies and mixing all the candies together. As steps 10–12 in chapter 4 ask you, in real life do you have an equal probability of sampling any individual in the species group of interest, regardless of that individual's particular species, sex, or age? If your answer is no, as it should be, how much confidence will you place in the accuracy of your quantitative measures of species diversity?

Even if these reflections leave you feeling depressed, DO NOT EAT ANY CANDIES YET.

much more diverse than Site 1? Naturally. Would you base a management decision on that thought? I trust you'd first examine the base of logarithms that was used in each case.

Many stories persist (and I don't doubt that some are true) from the days when environmental impact assessments were first required by law in the United States, and unscrupulous consulting firms hired by polluting industries exploited this unfortunate property of H'. For example, to demonstrate that the toxic discharge from an agrochemical factory has no negative impact on the watershed, one simply calculates the diversity of benthic invertebrates upstream from the plant using H' with base-10 logarithms and downstream using H' with natural logarithms (base e)—if any insects still survive downstream. Is there any chance this could happen today, in your own landscape? What if the firms involved hire unscrupulous or simply naïve environmental consulting firms to evaluate the environmental effects of commercial forestry, mining, road construction, or the discharge of waste into rivers? This possibility gives one more reason for preferring C_{inv} over H'. It's difficult to cheat with C_{inv}, since no logarithms are involved.

What's More Important, How Many Species or Who They Are?

Combining a list of values of n_i into a single statistic, like combining data into a single mean value, loses a tremendous amount of information. If you regarded all the species involved as interchangeable entities, this loss of information might not be serious. For example, an ecosystem ecologist investigating the relationship between species diversity and energy flow might reasonably justify her treatment of all the species within a given trophic level as equals. Or, for a basic study on the diversity of native frog species with respect to forest structure, H' or (better) C_{inv} might be adequate (although see below). If you're a conservation professional, though, many of your concerns about species diversity involve the concept of ecological integrity or the "health" of the native biota. Presumably, within a given group some species are more crucial than others to the biota's health or to ecological integrity as you've defined it. Presumably you'd consider an assemblage of plants or fishes that's composed of native species, for example, more "healthy" and more conducive to ecological integrity than an equally diverse plant or fish assemblage composed of aggressive, exotic invaders. Using a diversity measure without regard to who's involved may mislead you greatly.

Reexamine table 9.1. The species that contribute most to the high values of diversity indices in Site 3 are exotics; fewer native species exist, and in lower densities, than in Sites 1 and 2. Obviously, Site 3 is being, or has been, "wounded," and the suite of opportunistic plants has taken advantage of the wounds. In this case or many similar cases in real life, species diversity—as measured by H' or C_{inv}—may actually *increase* as ecological integrity is compromised, falsifying the fundamental assumption presented at the beginning of the chapter. Require me to increase the H' or C_{inv} of the herbs or earthworms in a forest fragment and I'll simply encourage all the exotics or pioneer species hanging around the edge to invade. Will you ever again use H' or C_{inv} alone to signal ecological integrity? I sincerely hope not.

Finally, by combining the sample data into a single index, you may also lose crucial information about the native species themselves. Sites 1 and 2 present nearly identical values for diversity indices, but the relative abundance of the different species differs greatly between the two. If certain of those species are of special interest, this change in ranking could be important. Now, wouldn't it be nice if you could condense all the information into a single format that would present all aspects of species diversity but would also tell you which species is which?

Representing Diversity with Rank-Abundance Graphs

In fact, a simple and nearly ideal alternative to species diversity indices does exist. This is not a single numerical value but rather a *rank-abundance graph* (also known as a dominance-diversity graph or "Whittaker curve").[3] To plot the rank-abundance graph for a sample of S species each having n_i individuals, first calculate each p_i value following equation 9.1. Then, calculate the logarithm (to the base 10) of each p_i value. Because raw values for p_i are all ≤ 1.0, the values of $\log_{10} p_i$ are all ≤ 0.0. Next, get a piece of graph paper and draw the two axes. The abscissa (x-axis) is "rank of species from most to least abundant" (from highest to lowest value of p_i or $\log_{10} p_i$). The ordinate (y-axis) is "$\log_{10} p_i$." Now, plot the log p_i value for each one of the S species, beginning with the most abundant one and ending up with the least abundant, often the species whose n_i is 1. Because points on the abscissa don't indicate any absolute value, simply the rank from most to least abundant species in the sample, on the same graph you can plot two or more samples that you wish to compare. In figure 9.1, I've plotted the data from all three sites in table 9.1. The final step may seem trivial but is critically important: *identify each point you've plotted with the name of the species it represents or a code for that name.* For practice, go to box 9.2.

Note that instead of values for $\log_{10} p_i$ you could plot values for $\log_{10} n_i$. Graphs using $\log_{10} n_i$ draw attention to disparities in N (sample size) between samples, a feature that may be helpful to your interpretation. The "tails" of all the samples, consisting of those species with $n_i = 1$, will line up horizontally. On the other hand, graphs using $\log_{10} p_i$ draw more attention to the comparative shapes of the different curves, to changes in the abundance rankings of different species, and to variation between samples in the relative amount of numerical dominance as opposed to equitability. Often, although not always, you may prefer graphs of $\log_{10} p_i$ to graphs of $\log_{10} n_i$. If you'd like to compare the two approaches, refer to table 9.1, construct the three rank-abundance curves but using $\log_{10} n_i$ instead of $\log_{10} p_i$, and compare your results with those displayed in figure 9.1.

Interpreting Rank-Abundance Graphs

With rank-abundance graphs you can compare all biologically important aspects of species diversity among samples. Each plot's span from left to right on the graph, of course, reflects the number of points it contains, or S. In figure 9.1, you see right away that the curve representing Site 1 has a broader span, therefore a greater S, than that representing Site 2. In the same instant you note,

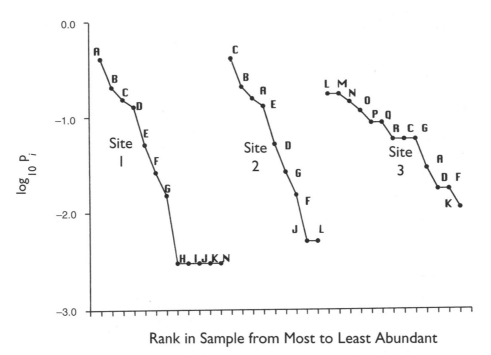

Figure 9.1.
Rank-abundance (dominance-diversity) graphs of the data in table 9.1. Note that the ordinate (*y*-axis) is on a logarithmic scale. The species for each sample are plotted from most to least abundant (highest to lowest p_i) within that sample. Several samples can be included on a single graph. Simply labeling the points shows which species is which, perhaps the most useful characteristic of these graphs.

Box 9.2. Drawing Rank-Abundance Graphs

Practice constructing rank-abundance graphs, either using values for the first sample of 20 candies you drew in the exercise of box 9.1 or using values from a new draw of 20. Follow the procedure under "Representing Diversity with Rank-Abundance Graphs" and illustrated in figure 9.1. Plot the $\log_{10} p_i$ for each candy "species" from the most to the least abundant. Points should be a standard distance apart on the *x*-axis. Label the top of the *y*-axis "0.0" and continue downward through −1.0 and, near the origin, −2.0 (in this case, you won't have any values lower than that). *Make sure you label each point with the corresponding flavor of candy (i.e., the species name).* Interpret this graph according to the guidelines in the text.

   Next, repeat the entire procedure, but this time based on a new sample of 50 candies randomly selected from the same bowl. Plot the new graph on the same pair of axes as the first one, as in figure 9.1. Does the shape of the curve change? Does the "tail" look different? Have species changed ranks?

   Now make up a second bowl (response unit) of candies with the following composition:

   70 candies of flavor (color) A

   100 candies of flavor (color) B

   1 candy of flavor (color) C

   1 candy of flavor (color) D

   5 candies of flavor (color) E

   30 candies of flavor (color) F

Box 9.2. Continued

Draw 50 candies from the new bowl and plot the rank-abundance graph on the same pair of axes you used for the two samples from the first bowl. Compare the three graphs. Does the shape change? Probably not much. Is the "tail" different? Probably not much, in shape at least. Have the species changed ranks? Of course. Reflect on this. Be patient and STILL DON'T EAT ANY CANDIES.

though, that this difference results from the different lengths of their "tails," the horizontal lines consisting of those species with $n_i = 1$ whose presence or absence may be sampling artifacts. Other than their tails, though, the shapes for Site 1 and Site 2 are identical, signifying that equitability or its converse, numerical dominance, is the same between the two sites. But wait: the sequence of species is quite different. Clearly, the three most common species are reversed between Sites 1 and 2. If species C, or A, is of special interest, then this information may be very important.

The greatest visual contrast between graphs, though, is that between Sites 1 and 2 on the one hand and Site 3 on the other. The shallow slope of the curve for Site 3 signals its greater equitability among species, with several species of moderate abundance and none with such pronounced numerical dominance as species A at Site 1 or species C at Site 2. Who makes up that diverse assemblage in Site 3, though? A quick glance at the labels tells you that Site 3 is dominated by exotic species. In fact, the graphs point out that those native species dominant in Sites 1 and 2 are relatively rare or else (species B) missing entirely. Of course, you could obtain all the same information by a prolonged scrutiny of table 9.1. The graphs, though, present the information in a much more striking and accessible form.

Complementing Graphs with Indices

Rank-abundance graphs may satisfy your desire to have all the information important to conservation concerns, and to many questions in basic ecology, presented all at once. They won't satisfy your craving for a numerical measure, though. You might decide to complement graphs with diversity indices calculated after you've scrutinized the data in the graph or the original table. For example, calculate C_{inv} for native species (with N now the total for the n_is of those species only, and p_i recalculated accordingly) or for a certain guild of special concern. There's nothing intrinsically wrong with a numerical species diversity index, as long as you don't rely on it exclusively. I strongly suggest, though, that you use rank-abundance graphs as the primary approach to evaluating the species diversity of an indicator or target group for purposes of conservation, management, and basic ecology as well. Later in this chapter, I'll emphasize this point even more strongly.

The Scale of Species Diversity

A presentation of species diversity, whether by indices or graphs, only applies to the time frame and spatial scale of the evaluation unit you used. Expand the size of the evaluation unit either in time or space, and you'll soon include additional species (Halffter 1998). As chapters 4 and 6 suggest, move the evaluation unit to a different time or space even within the same response unit, and you'll probably encounter different species or at least change the p_i values for species already on the list.

This feature, heterogeneity in species composition from site to site, provides a very useful, different way of viewing species diversity. Until now we've considered the diversity of species within a certain region, small or large, of the landscape. We label the term diversity at these two different scales as *within-habitat diversity* and *regional diversity* (sometimes called α- and γ-diversity), respectively. Return once more to figure 4.1. The species diversity of forest frogs within a given SL or UL plot would be within-habitat diversity, whereas their diversity within the reserve as a whole, including all three vegetation types, would be regional diversity for this particular case. How do we get from one to the other, and why should we bother?

Let's define *between-habitat diversity*, or β-diversity, for a given group as the amount of change in its species composition from one part of the landscape to another (e.g., see Halffter 1998). A landscape that has exactly the same species composition throughout presents a β-diversity of zero by definition. Local (α) and regional (γ) diversity are one and the same. A landscape across which species composition changes rapidly presents a high β-diversity by definition. In this case, the landscape as a whole supports many more species (γ-diversity) than any single habitat within it (α-diversity).

For several decades ecologists and conservation scientists have focused most of their attention on within-habitat (α) and regional (γ) diversity. Few indices exist for representing between-habitat diversity—a fact for which we should be grateful. A simple measure might be the rate of change in species composition per unit distance. For example, you might go from one end to another of the reserve in figure 4.1, stopping every 1,000 m to run frog transects for three consecutive nights. By comparing the relative similarity, or difference, in species composition between each successive pair of samples (box 9.3), you'd discover where the greatest changes in species composition occurred per

---

Box 9.3. Calculating the Similarity between Samples

Many numerical techniques exist to quantify the similarity, or difference, in species composition between samples. Several forms of multivariate statistics (see chapter 5) are just sophisticated ways to compare many samples at once based on their species composition. For comparing two samples at a time, though, many ecologists return to an extraordinarily simple, straightforward measure: *proportional similarity*, or *PS*. To calculate the *PS*, examine the raw (not logarithmic) values of p_i for the species in sample 1 and sample 2. For each species i that occurs in either sample or both, note the lesser of the two p_i values, that is, p_{i1} or p_{i2}. If the species is in one sample but not the other, the lesser of the two values is, of course, zero. Add up the values you've recorded. In other words,

$$PS = \sum_{i=1}^{S} \min(p_{i1}, p_{i2}) \tag{9.5}$$

The value of the *PS* ranges from 0 (no species in common between the samples) to 1.0 (both samples with identical species in identical proportions). For example, the *PS* between Site 1 and Site 2 (table 9.1) is 0.154 (Species A) + 0.205 (B) + 0.153 (C) + 0.051 (D) + 0.051 (E) +

Box 9.3. Continued

0.015 (F) + 0.015 (G) + 0 (H) + 0 (I) + 0.003 (J) + 0 (K) + 0 (L) = 0.647. In other words, Sites 1 and 2, or, better said, the two samples from Sites 1 and 2, are 64.7 percent similar to one another. Likewise, the *PS* between samples 1 and 3 is 0.139 and that between samples 2 and 3 is 0.146. Check my calculations.

Now, return to the candies. (1) Calculate the true *PS* between the two bowls of 207 candies each (response units) that you set up according to the recipes in boxes 9.1 and 9.2. Of course, as a field investigator you'd be unable to do that. (2) Now calculate the *PS* between the two *samples* of 50 candies each, one from each bowl, that you used for constructing the second and third rank-abundance graphs in box 9.2. Does the sample *PS* match the true value? Is it smaller or larger? (3) Finally, draw four random samples of 15 candies each as follows: samples 1 and 2 from the first bowl, replacing the candies after each draw, and samples 3 and 4 from the second bowl. Calculate the *PS* for each of the six possible pairs: 1–2, 1–3, 1–4, 2–3, 2–4, and 3–4. If your samples were perfect mirrors of the underlying candy assemblage, the *PS* between any two samples drawn from the same response unit—such as the *PS* between samples 1 and 2, or that between samples 3 and 4—would always be 1.0, right? What values did you actually get for the comparisons 1–2 and 3–4? And if these samples of 15 candies each were perfect mirrors of the species compositions of the response units (bowls) from which they were drawn, what should the *PS* be between samples 1 and 3, or 2 and 4, or 2 and 3, or 1 and 4? What values did you actually get for those comparisons? So is the *PS* a "true" measure in which you can have absolute confidence, or is it just one more sample statistic with variable accuracy? How will you apply this newly acquired wisdom to designing and interpreting a real-life inquiry that compares different samples based on their species composition?

Note that by definition β-diversity (see text) increases with increasing *dissimilarity*, not similarity, between successive samples along a gradient. So, if you wish to apply this technique to questions about β-diversity, simply calculate the *proportional dissimilarity* or *PD* as

$$PD = 1 - PS \qquad\qquad (9.6)$$

That is, the greater dissimilarity between species assemblages from point to point across the landscape, the greater the β-diversity. Note that Colwell and Coddington (1994) use the term *complementarity* in place of dissimilarity.

Yes, NOW you can eat the candies.

1,000 m traversed, the sites of highest β-diversity. From the point of view of frogs, those are the sites where different sorts of habitats are packed in most closely to one another, where numerous sources and sinks are jumbled together.

Why worry about β-diversity at all? Isn't it rather esoteric? Consider the definition of conservation given in chapter 1, and the suggestions at the end of chapter 6. If you had to choose sites for intensive conservation efforts, which sites should concern you more, those with high β-diversity or those with low β-diversity? Which will give native species the greatest number of options for surviving during episodes of climate change? In which will the fauna likely be most sensitive to stresses? At least one international conservation organization is now identifying "biodiversity hot spots" based on β-diversity, rather than the traditional α- or γ-diversity (also see Poiani et al. 2000).

What to Do?

Once again, ideally you should approach biodiversity conservation on a number of different levels (Noss 1990), from genetic diversity within populations to the diversity of ecological processes. In practice it's often easier to tally species than genes or ecological processes. Whatever the case, when you're dealing with the concept of species diversity:

1. Focus on α-diversity. Choose the evaluation unit and response unit with extreme care. Are you confident you're sampling a reasonable proportion of the underlying species assemblage, or is your sampling so weak that it's scarcely better than drawing two candies from a bowl of 207 (box 9.1)? After sampling, construct and scrutinize rank-abundance graphs. If you absolutely insist, supplement the graphs with diversity indices or simply S, but be careful of which species you include in the index (see above). In theory, the best indices to use would be those proposed by Hurlbert (1971). Unfortunately, if you used those, few of your colleagues would understand what you were talking about. So if you must, use the better of the two widely recognized alternatives, C_{inv}. Never apply H' unless you have a very good reason for preferring it to C_{inv}. If you must use H', always report the base of logarithms you used.

   Of course, you could also apply a similar approach (rank-abundance curves, supplemented if absolutely necessary by diversity indices) to other levels of biodiversity. For example, you could examine the diversity of habitats within a landscape, with p_i now representing the proportion of total area occupied by habitat type i. Or you could examine the diversity of genotypes within a plant or animal population, with p_i the proportion of individuals having Genotype i.

   Especially in Latin America, the use of species diversity indices among both basic ecologists and conservation professionals is widespread and increasing exponentially. I can't help but state an opinion that's even stronger than those you've encountered elsewhere in this book. With respect to biodiversity conservation, *the burden of proof is on those who use, or would use, species diversity indices, to justify unequivocally and explicitly why the numerical indices are appropriate and necessary to the goals of the inquiry or the presentation afterward.* To tell the truth, I cannot think of one single case where the all-in-one species diversity indices would be necessary, appropriate, or even helpful to biodiversity conservation per se. Furthermore, having used species diversity indices in my own work since 1971, upon reflection I am now skeptical about their value even to studies in basic ecology, except for very specific purposes such as expressing the diversity of food resources from the viewpoint of a particular group of animals. In general use, the indices simply lose far too much information that's biologically significant. They're far too easily misused and misinterpreted. The numerical value that an index yields does not tell us anything that's biologically meaningful unless we're still trying to demonstrate a relationship in the field between species diversity and ecosystem stability, an approach that was quite thoroughly discredited by the late 1970s.

2. Consider β-diversity. Does the rate of change in species composition appeal to you as a broad-scale measure of biodiversity that has important conservation implications? If so, consider some inquiries into β-diversity. You could begin by employing the simple technique presented in box

9.3 to examine samples from transects along climatic gradients, along altitudinal gradients, or perpendicular to major features of topography. As in investigating α-diversity (table 9.1), though, be careful whom you include in this measure. It's much more important to apply common sense and an appreciation of natural history to questions of species diversity than to apply sophisticated mathematical skills. Keep in mind that species aren't just numbers.

Extending the Reach of Inquiry

Ultimately, conservation depends on people who are neither biologists nor managers.

—Anonymous

Are conservation professionals and field biologists the only ones capable of understanding and applying scientific inquiry to questions that intrigue or worry them? The example of the campesino in chapter 2 suggests otherwise. Biologists and other conservation professionals scarcely have a monopoly on conservation. To expand on the quote above, conservation for the long term resides in the hands, hearts, and minds of people who will be around long after the last reserve manager, biologist, or other conservation professional and barbed-wire fence around the protected area has faded from memory. Therefore, today's conservation strategies must include plans to empower a variety of people to make thoughtful decisions among the alternative choices of dealing with their natural surroundings—even though those decisions might not be those we'd necessarily make ourselves. For many pages I've discussed the inquiry process, common sense, and natural history knowledge as critically important means to empower you, as a conservation professional or field ecologist, to make the best possible decisions. Might those tools help to empower other people as well?

This chapter suggests a few ways in which scientific inquiry might extend to several groups of people, all involved in the fate of protected areas or larger landscapes yet few having any intention of becoming card-carrying conservation professionals or ecologists themselves: local communities in the semi-natural matrix that surround the protected area include adults, or those people currently making decisions affecting conservation in that very area, and children, or those who will be making such decisions a decade or so from now. Park guards, often from local communities, patrol for

145

violations of the protected area, thereby spending a great deal of time in the field and sharpening their powers of observation. Visitors of all ages come from landscapes near and far to wander through the protected area's officially designated "visitor center" or to stroll along its trails before returning home and continuing to make decisions affecting conservation in their own day-to-day surroundings.

You've probably seen many efforts to sensitize these various groups to conservation issues. For example, initiatives in participatory management or community-based conservation already involve local people near some protected areas.[1] Programs in environmental education target schoolchildren in cities and rural areas alike. Park guards frequently attend workshops on the principles of biology and ecology. "Nature interpretation" and "nature interpreters" introduce visitors to the local biota and ecological processes. Can you contribute anything new to these efforts? Absolutely, and it's critical that you or those working with you do so. Few current approaches challenge local community members, schoolchildren, park guards, or visitors to ask questions about their surroundings, and fewer still require those people to answer their own questions through firsthand experience and then reflect on the answers. Now that you yourself are adept at scientific inquiry, consider sharing—but never imposing—your expertise.

Local Communities and Scientific Inquiry: Participatory Conservation

Ideally, any conservation strategy would involve collaboration with local people at every stage of the management cycle (see Margoluis and Salafsky 1998), whether the concern that begins the cycle (figure 2.4) is one brought up by the community members themselves or one that you bring to their attention. Local people may be recently arrived colonists, indigenous communities with centuries of history in the region, or a mixture. These people have natural feelings of ownership not only toward the semi-natural matrix in which they live and work but also toward the protected area nearby, which they probably know at least as well as you do. Such proprietary sentiments, which often frustrate conservation efforts, might instead be turned to conservation's benefit if community members know that full-scale inquiries are addressing concerns and questions they've generated themselves. How might you facilitate this? It's not enough just to hold a meeting with the community, ask for their concerns or questions, and promise to take them into consideration. You must provide a context for those questions and go on from there. What's the context? The inquiry cycle. The following is one possible scenario.

Suggestions for Workshops to Initiate "Community Inquiry"

Collaborate with community leaders to organize a one- or (preferably) two-day workshop as a joint venture. If local customs permit, suggest that all the community's adults, men and women alike, be invited. Schedule the workshop for the full day, from as early in the morning as convenient until the hour for household chores in late afternoon. Offer to provide or at least subsidize the midday meal (food and drink) as well as refreshments for midmorning and midafternoon breaks. Provide a notebook and pencil for each participant. Ask that the workshop be held in and around the school building or other community center.

Participants in the workshop should include not only interested community members but also a small number of "facilitators," including yourself. The facilitators might include park guards or biologists, all of whom have already practiced the inquiry approach (see below). The workshop might proceed as follows:

1. As the leader, introduce the goals of the workshop. The fundamental goal is to empower the participants in the approach of scientific inquiry. One practical goal is to encourage them to apply that approach to immediate concerns about the fate of the local landscape. Another practical goal is that they participate fully not only in getting answers to their questions but also in applying those answers. The local landscape includes not only the semi-natural matrix in which they reside but also the protected area nearby, if any.

2. Ask all participants to introduce themselves.

3. As the leader, present the underlying theme: conservation and sustainable use of the local landscape and its resources over the long term. Begin by asking the participants two loaded questions: "Do you want your children and grandchildren and great-grandchildren to be able to live in the same landscape that surrounds you today? Do you want them to have access to the sorts of natural resources that you yourselves use today—fertile soils, wild game to hunt, fish in the rivers, medicinal and other useful plants in the forest, clean water?" Most likely, participants will answer with a resounding yes to both questions. Point out that what they're saying is that they wish to conserve the local landscape and its important elements. Very briefly, emphasize that conservation isn't by any means just a top-down policy of setting aside protected areas and restricting their use. Rather, it's a means of ensuring that the landscape and its natural resources are available to future generations. Ask, "If that's the case, then how can we accomplish conservation?" Present the scheme of figure 1.1, modified as desired, to justify the need for inquiry (investigation) in achieving conservation goals.

4. Discuss "how to become a scientific inquirer." Present the inquiry cycle (figure 2.2). Illustrate the inquiry cycle with examples from the participants' daily lives, such as the example of the campesino in chapter 2 or another that's more appropriate to local conditions. Emphasize that the first, and most important, step of the inquiry process is to ask questions.

5. Practice asking questions. Everyone goes outside the building and repeats what you yourself did in box 2.1. You assign to each participant a different, unique patch of ground, about 50 × 50 cm, that presents some small-scale heterogeneity. All then draw crude maps of their mini-landscapes and come up with at least five questions about what they observe within the confines of the parcel. There are no requirements whatsoever as to the nature or format of the questions (figure 10.1). Emphasize that there's no such thing as a stupid question, as long as it ends in a question mark. Depending on social dynamics and degree of literacy, participants might work in pairs instead of singly, or even as teams, each with one scribe. Facilitators must take care here and elsewhere not to impose their own perspectives.

6. Reconvene. Remind everyone that there's no such thing as a stupid question, then ask each participant or team to read off one or two favorite questions from the list. You may be impressed by the observation skills evident in the questions. Be positive and encouraging—don't voice even constructive criticisms of these questions.

7. Return to the four steps of the inquiry cycle (figure 2.2). Point out that the question is the first and most important of these steps, but that the other steps must also take place if there's to be an answer and if that answer is to be useful. Now assert, "Yes, all questions are valuable but

Figure 10.1.
Practice with constructing questions.
Members of the community of San
Antonio de Cuellaje, Ecuador
(Imbabura Province), asking questions
about the mini-landscapes they've been
assigned. Photograph by Marty Crump.

some may be more valuable than others with respect to their ability to lead smoothly into the action, reflection, and application steps." Present the four criteria for questions (in chapter 3) one by one. Discuss and illustrate means of converting questions that don't conform with the criteria to ones that do.

8. Assign participants to teams, each composed of three or four community members and one facilitator. If the workshop is for two days, these teams will first practice formulating questions that comply with the four criteria. The members of each team wander about within a 50-m radius of the central location. They observe, come up with questions whose spatial scale is at most that 100-m-diameter circle, negotiate among themselves, select one question, and then refine it until they believe that it complies fully with the criteria. Once all teams are ready, each presents its question verbally. In each case the listeners ask themselves, "Does the question comply with the four criteria, and if not, how can the wording be changed so that it does comply?" Guide this phase of constructive criticism as appropriate. If the workshop is for two days or more, as time permits have yet another practice session, but first make up new teams. On the other hand, if the workshop lasts one day only, you'll need to skip this stage of practice entirely—though not the first practice of the 50×50 cm parcels—and proceed directly to the next step.

9. Make up new teams with at most one outside facilitator per team. Ask each team to elect one of its community members as the spokesperson. Each team's independent task is to articulate a real-life question that springs from local management or conservation concerns—or similar concerns, such as those about agroecology—and that can be answered through scientific inquiry. That is, the teams wander about independently, observe the landscape near and far, ponder, discuss concerns that can be addressed with inquiries, propose questions for those that comply with the four guidelines (chapter 3), negotiate, and then propose and refine one question per team (figure 10.2). You yourself did something similar (box 3.2).

If the idea of the workshop is to address a conservation concern that you yourself have brought to the workshop (figure 2.4), then ask the participants to direct their questions toward that theme. Examples of such themes might be harvest of nontimber forest products, hunting, selective logging, small-scale mining, edge effects, or watershed protection. If the idea of the

Figure 10.2.
Generating real-life questions. Residents of the buffer zone for Bwindi Impenetrable National Park, Uganda (Rutugunda Parish), debating questions about the sustainable use of medicinal and basketry plants in the outer reaches of the park.

workshop is to address concerns generated by the community members themselves, point out that the members of each team should agree on the concern they wish to address before constructing the question. A few examples of the myriad additional concerns that local people might have include the control of insect pests, management of crops or crop parcels, pasture management, crop raiding by mammals or birds sallying forth from a protected area, other possible effects of having a protected area nearby, and water quality (see also Castillo and Toledo 2000).

10. Reconvene the entire group. Each team presents its question. Now encourage full-scale constructive criticism, discussion, and debate on all aspects of each question proposed.

11. Vote (community members exclusively) on the questions, ranking these from most to least popular.

12. Now bring up the next critical phase, the arrow that connects the question with the action step—that is, the way to design the inquiry that will best answer the question. If you have two or more days available, first go through one or both acts of appendix D; that is, familiarize the participants with the fundamentals of design and statistics.

13. Guide the group in roughing out the outline of a study design for the top-ranked question. Discuss the nature of the results that the study might produce and how, after the step of reflection, those might be applied to addressing the original concern.

14. If time permits, repeat the process for the second-ranked question. Make sure you've reserved time for the last steps, though.

15. Discuss the means of collaborating throughout the inquiry process to follow. As a group, decide on the responsibilities and commitments of the interested local people, yourself, other biologists and conservation professionals, and other institutions. Take special care that the full responsibility for the complete design, the field work, the analysis, and the reflection doesn't gradually shift back to you and other professionals. Instead, take steps to ensure that interested community members will continue to participate fully, on both intellectual and practical aspects of the inquiry.

16. Elect a committee consisting of half local people and half facilitators. Include yourself if the question the community has selected is within your area of competence. If it's clear that the inquiry will benefit from the advice of other outside people, the first task of the committee will be to contact such individuals.

17. Bring the workshop to a close, on a high note. Reemphasize the commonsense nature of the inquiry process, design, and real-life application. Reemphasize the collective nature of the enterprise.

18. Get to work!

Is "Community Inquiry" a Pipe Dream?

You may be amazed, as I've been, at the speed with which some members of some communities absorb the fundamental themes of this book and apply them in novel ways. By the end of a workshop, with some guidance on the facilitators' part, the community will have proposed, debated, refined, and ranked several questions, all conforming to the criteria of chapter 3 and all addressing real-life concerns, yours or theirs. Ideally, you and they will also have begun to rough out the designs for the inquiries themselves and will have assigned responsibilities for the various steps. Even if only a few local people participate fully in the study that follows, you can continue to involve the community as a whole. Schedule an evening workshop once a month at which those directly involved in the study discuss progress, describe setbacks and discoveries, present preliminary results, and ask for input. Once all the data are in, fully involve the community once more to help evaluate the results, reflect, propose conservation or management guidelines, apply them, and if possible monitor the consequences. If you succeed in this intellectual and practical partnership with local people from beginning to end, at least some may feel "pride of ownership" in any guidelines that result and will take it upon themselves to make them work.

Realistically speaking, will community inquiry succeed? Sometimes but not always. Not all biologists or conservation professionals will be willing to delegate so much intellectual responsibility to local people or to allow the questions that local people generate to guide the nature of the research to follow. These professionals may subtly change the question into one that fits their own interests. The community side isn't always harmonious either. The "noble community" is no less a myth than the "noble savage." Every community has hidden—or not so hidden—undercurrents of internal problems, strife within and between families, political machinations, and serious feuds. Sometimes by working with person A, you're automatically on the blacklist of person B. The protected area, its management, and its managers may cause resentment, particularly among recently arrived colonists who, unlike some indigenous groups, have been granted no rights to the use of the protected area's resources. Particularly conflictive situations exist where collaboration of any form between the community and conservation professionals is unthinkable at present. Nevertheless, "community inquiry" and true collaboration can succeed and are succeeding in some cases. Try it.

Local Children and Scientific Inquiry

Today's children are the decision makers of tomorrow. It would be ideal if the next generation were already gaining experience at asking and answering questions about the surroundings they encounter daily, such that on reaching adulthood they'd already be objective thinkers, accomplished inquirers, and conscientious decision makers on how best to conserve and utilize local biodiversity. Is this lofty goal distant from your responsibilities and your expertise? Not at all. Consider introducing the inquiry process (figure 2.2) into local schools as science education and environmental education. You'll find that many rural schoolteachers enthusiastically adopt the concept of intellectually challenging, low-cost scientific inquiry that takes place on the school grounds, especially once they realize that this approach can end up covering a considerable part of the mandated curriculum. In short, consider ways of encouraging the development of local programs in schoolyard ecology (Feinsinger, Margutti, and Oviedo 1997; Rozzi, Feinsinger, and Riveros 1997; Arango, Chaves, and Feinsinger 2001).

The first step might be a training workshop for elementary school teachers so that they, and later their students, become adept at doing concise but complete scientific inquiries involving the animals, plants, interactions, and consequences of human activities on the school grounds. Schoolyard inquiries, like those that you carry out, involve not only the question and action steps of the inquiry cycle but also the reflection step. In the upper elementary grades at least, children will learn to reflect on their findings at the small scale of firsthand experiences in the schoolyard and to relate these findings to phenomena at the larger scale of the surrounding countryside, the scale at which they'll be making decisions later on in life. Appendix C provides sources for information on schoolyard ecology.

Alternatively, if you work within a protected area, groups from local schools may pay *you* visits. When school groups arrive at the protected area, don't just lecture to them or encourage them to wander aimlessly through the visitors' center and along trails. Instead, personally guide the schoolchildren and their teachers through rapid but complete, age-appropriate, hands-on inquiries you've thought up, ideally in collaboration with the teachers. Here, too, focus on insects and spiders, plants, microhabitats, ecological processes, and the various "footprints of human beings." Guide the reflections that follow toward jargon-free discussion of the advantages and disadvantages of different conservation alternatives, land uses, and kinds or levels of protection. Invite the children to discuss your real management concerns, based on the firsthand experience they've just had. As you guide the students through the reflection step, at least the older children may begin to comprehend the rationale for a protected area, the complexity of its management, its relationship with the semi-natural matrix, and the different perspectives of different stakeholders. Your small investment of time may pay off many times over in the future.

To date, visionary and enthusiastic local people—ecologists, university students, conservation professionals, education professionals—have instituted schoolyard ecology at sites in Mexico, Belize, Venezuela, Colombia, Ecuador, Peru, Bolivia, Argentina, Chile, and Brazil. While these sites include large cities, many of the communities already practicing schoolyard ecology reside in the buffer zones of, or inside of, protected areas. Interested? If so, return to appendix C.

Park Guards and Scientific Inquiry

Many guards in national parks or other protected areas have one assigned task and one only: patrolling for violators and violations of park policy. The task is rarely intellectually stimulating, though often it is dangerous and always exhausting. Consider, though, that park guards almost always come from rural backgrounds, often from communities near the protected area. While patrolling, they're constantly in the field, observing all that's going on around them not only with respect to violators but also with respect to the natural history of plants, animals, species interactions,

Figure 10.3.
Conservation professionals (guards and other personnel of
national parks in southern Argentina) observing the landscape
and framing questions (San Carlos de Bariloche,
Río Negro Province).

and other ecological processes as well as to potential threats to ecological integrity. Most park guards are already curious about what they see and are already framing questions (figure 10.3).

If you're not a park guard but you work with them, consider encouraging and tapping this tremendously important human resource. First, urge park guards to get involved in community workshops (see above), schoolyard ecology workshops, and initiatives that will ideally follow each of those. Second, if possible encourage park guards to undertake their own research in the field. You can easily provide them with the tools: the management cycle (figure 2.4), the field research cycle (figure 2.5), and the fundamentals of design (appendix D).

South American colleagues and I are currently trying out means to provide park guards with these tools. For example, the guards are invited to be facilitators in four- or five-day workshops in schoolyard ecology, where at the same time they become familiar with the inquiry cycle and the four criteria for research questions. Then the guards stay on for a second workshop that lasts two to four days. Here we introduce the more complete versions of the inquiry cycle (figures 2.4, 2.5) and then the fundamentals of design and statistical inference (appendix D). We work through the example of the selected logging question (chapter 4). Next comes the discussion of some ten or twelve of the most critical steps of the design process (chapter 4), using accessible language. Concentrating on the immediate surroundings of the workshop site, teams next practice framing questions of interest to them and designing the studies to answer those questions. Finally comes the real thing. Each participant, or group of participants from a given center, develops the question and preliminary design for a study they're truly interested in carrying out once they return to their post.

Given the expanses of habitat they cover during patrols and the repeated visits they make to any one site during the course of a year or more, park guards are in a unique position for carrying out long-term, well-designed, landscape-scale inquiries on questions of biodiversity conservation or basic natural history. Little or no additional time need be invested: they can record data while patrolling. In addition to fostering the intellectual growth and self-confidence of the guards themselves, such inquiries could be of immense value to the protected area's management and to conservation science, or basic field ecology, at large. For those guards who initiate studies upon their return to work, we offer to provide advice and feedback as needed. In some cases we hold a competition later on for the best studies. Winners are encouraged, and offered support when possible, to present their studies at national or international congresses. You can probably think of many additional ways to involve park guards in scientific inquiry.

Visitors and Scientific Inquiry

This book's main theme is the role that scientific inquiry can play in making decisions related in some way to biodiversity conservation. There's one other group of people who make such decisions: visitors to the protected area. Of course, such visitors rarely make decisions that have to do with conservation of the protected area per se. Instead, their role in biodiversity conservation, positive or negative or neutral, is played out in their home landscapes. That is, upon returning home, they'll continue to make decisions on issues that may directly or indirectly affect the fate of the biodiversity of their own surroundings. Perhaps visitors' experiences in your landscape can influence the nature of those decisions. With this scenario in mind, how might you enhance those experiences? By exposing visitors to short, snappy firsthand inquiries that lead them to reflect on what goes on in their very different day-to-day surroundings (see box 10.1).

Box 10.1. Points of View: An Inquiry for Visitors

This inquiry normally calls for six people besides the leader, but it can be adapted for a group of any size. Although it's most appropriate for a guided group, it could also be modified for self-guided visitors, by means of a placard to be posted along the trail. The inquiry is a bit unusual in that it doesn't involve any direct contact with the organisms concerned; on the other hand, it works just about anywhere there's a good view of the surrounding landscape, and it's entirely nonintrusive.

Question: Do different animals view the same landscape in different ways?

Rationale: Our own point of view as human beings resides about 1.0–1.8 meters above the ground and has a radius of about 50 m. Might animals having other sizes, shapes, and modes of travel have other points of view, and if so what might the consequences be?

Methodology: For a group of six, the leader inserts a twig or toothpick in the ground at a point that presents some heterogeneity at a very small scale, for example, leaf litter, dead grass stems, sand grains, pebbles, and plant seedlings. Each participant takes the point of view of a different animal, as specified below, and sketches a rough map of the way the landscape looks from that point of view.

The animals and their scales are: a flea or mite, whose entire landscape is the square 2 × 2 cm centered on the twig; an ant, 20 × 20 cm; a mouse, 2 × 2 m; a sparrow, 20 × 20 m; a

Box 10.1. Continued

pigeon, jay, or other moderately large, fast-flying bird, 200×200 m; a condor or eagle, 2×2 kilometers. The leader reiterates the rules: (1) all the maps are drawn at the same size (about 20×20 cm on a sheet of paper); (2) the twig is at the center of every animal's landscape; (3) all maps are drawn from above looking straight down as if they were maps in an atlas; and (4) each animal can recognize at most five distinct *classes* of landscape elements at his or her scale, although there could be multiple examples of any one class.

After spending about ten minutes completing their maps, the participants display and describe them one by one to the group until all six are lined up in a row, from flea to condor. The leader guides the discussion and the reflection.

Some prompts for reflection: In what ways do the predominant traits of the landscape change from one animal to the next? Does the flea perceive some of the elements that dominate the landscapes of the other animals? Which? Does the condor? Which? What's the nature of the most obvious "borders" between different landscape elements at each one of the scales? What shape are those borders? Do they represent elements created by humans or elements not created by humans?

Starting with the flea and going up, at which scale do elements attributable to human beings first appear as the landscape's predominant features? Does that scale relate to the horizontal scale (the 50-m radius) of our point of view? Just looking at the maps, are there any particular shapes typically associated with those elements? Continuing upward in scale, is a point reached where elements not created by humans begin to intrude once again into the landscape, even to dominate it? What sorts of features are those?

How might the basic nature of each of the six maps have changed during the past one hundred years? Does the landscape perceived by the flea of today resemble the landscape its great-great-great-great-grandmother flea perceived in the year 1901? What about today's sparrow versus its ancestor in 1901? Today's condor versus its ancestor in 1901? How could you determine that? Perhaps by consulting old maps and archives? Or by interviewing elderly residents? How do the activities of people today continue to affect the landscape at each one of the scales?

Now think back on the landscape where you yourself live most of the time. At each one of the six spatial scales, what might be the similarities and differences between your home landscape and the one in which you're standing right now? Which landscape displays the heavier "footprints" of human beings? Which of the two landscapes pleases you more aesthetically? Which of the two is the more practical for living and working? In general, where would you truly prefer to live, in a "pristine" wilderness landscape, in a landscape that displays a mixture of "natural" and human elements at various scales, or in a landscape dominated by human-engineered elements at every scale? What would you like the future of your own landscape to be, at each of the various scales or at least those of the sparrow, pigeon, and condor? What decisions commonly made by you, your family, and your friends might have some influence on that future? How might you modify those decisions such that there's a greater probability that your landscape will attain the future state you'd like the most?

You can expose visitors to the inquiry approach in at least three ways (Feinsinger, Margutti, and Oviedo 1997). (1) Sometimes park personnel themselves lead groups of visitors on guided tours. Every so often the leader might pause and guide the group through a rapid firsthand inquiry on the spot. Box 10.1 presents an example of such an inquiry, one that manages to touch on a fair number of the themes in chapter 6. (2) Many protected areas have a visitors' center or are designing one. Instead of providing visitors with passive information, some displays in the center could entice visitors into going through a very simple but complete inquiry, for example, based simply on a comparison of two different photos. Even better, the grounds surrounding the center—even the grass lawn or the dirt parking lot—could have signposts that entice the visitor into going through a rapid firsthand inquiry. (3) Almost any protected area has trails for visitors. In some cases, at least one trail close to the visitors' center already functions as a "self-guided nature trail" with informational signposts. These could be transformed into placards that guide the visitor through a complete inquiry at the spot, instead of simply providing the visitor with information that will be received passively if at all. Each placard could present a question and its context in brief, suggest how the visitor might answer the question by looking or listening or smelling or touching, and then present a prompt for reflection. The last of the prompts could lead the visitor to reflect on his or her home landscape, as in box 10.1.

It's unrealistic to hope that all visitors will take the time to go through inquiries no matter how brief they are. After all, unlike schoolchildren, visitors are under no obligation to learn. Still, if your efforts reach just 5 percent of the visitors, and if just 5 percent of those end up making decisions more conscientiously upon returning home, your efforts have been worthwhile. If you're interested in the possibilities of applying the inquiry approach to the audience of visitors, turn once more to appendix C.

Field Biologists and Conservation Professionals as Educators

Extending the reach of inquiry to local communities, to local schools, to visitors is public education in the broadest possible sense. Probably you don't have formal training as an educator. Does this mean that you're not permitted to engage in education? Certainly not. Conservation, scientific inquiry, and education are firmly linked. Farsighted conservation strategies, and farsighted conservation professionals and field biologists, will recognize the links and will realize that education of many different publics is the best hope for biodiversity conservation. True education doesn't consist of preaching, though. The field biologist or other conservation professional acting as an Expert, telling people what's Good and what's Bad, what they Should and Shouldn't do to "Save the Earth" and "Be Ecological," is not education. It's an attempt at indoctrination, an approach that could easily backfire and have the opposite effect to what was intended (Sobel 1995). Instead, education is empowering people to observe, to question, to reflect—in short, to think on their own. Could this form of education, and the inquiry approach itself, be extended to another group of people who exert tremendous influences on biodiversity conservation—that is, politicians and policy makers? Perhaps. I'll leave it to you to think about ways that might be done.

By working with groups of people outside the professional circles of conservation and biology, we're not only engaging in "inquiry outreach" but also opening ourselves to "idea input." The conservation concerns and inquiry questions that local people will bring to our attention cannot help but strengthen our own conservation perspectives, inquiries, and strategies (see also Margoluis and Salafsky 1998). We mustn't take ourselves so seriously. Conservation professionals alone don't have all the answers. We don't have all the questions, either.

Calculating Confidence Limits for the Population Mean

The goal of calculating *confidence limits* is to infer or estimate the value of a given parameter for the statistical population from which you've just sampled. Let's consider confidence limits for the unknown population mean μ. You can never state the exact value μ with 100 percent confidence unless you've actually examined that entire statistical population. With your sample data and some basic assumptions, though, you can calculate a range of values or a *confidence interval* that, following the working definition, will probably include the unknown value of μ, P percent of the time. The complete definition is somewhat more complex (chapter 5, note 6), but, pragmatically speaking, P estimates the probability that the lower end of the interval you calculate is less than μ, and that the higher end is greater than μ. The mathematical expression for the 90 percent confidence limits, for example, is

$$P\{[\text{lower value}] \leq \mu \leq [\text{upper value}]\} = 0.90$$

This assumes that (a) you're sampling randomly from the statistical population in whose μ you're interested, (b) each observation is independent from the others, and (c) the values in the statistical population itself display what's called a "normal distribution."

Before getting into the practical details, let's consider the complementary concepts of the precision of your estimate, and your confidence in it. The narrower the confidence interval, by definition the more precise the estimate (see Zar 1999). The higher the value of P, the greater your confidence by definition (that is, the lower your risk of being wrong by stating that the interval really does cover μ. Ideally you'd want to be precise and confident at the same time. As you'll see, though, with a limited sample size you must choose between two options: greater precision with less confidence and greater confidence with less precision. Which one you choose depends on the conservation concern

157

or study goal, your question, and the costs (speaking most generally) of being imprecise on the one hand or wrong on the other.

For a sample of size n, to calculate the confidence limits of μ you need the following: $\bar{x}$, $SE_{\bar{x}}$ (equation 5.8), your desired level of confidence P, and table A.1. Table A.1 presents values of a statistic called t for each combination of two other statistical variables whose definitions don't concern us just yet: α and *degrees of freedom* (often abbreviated as df, as in table A.1 here, or as v in some texts). Turn to table A.1 and look for the intersection of the column headed by that value for α that corresponds to the quantity $(1 - P)$, and the row for that value of df (degrees of freedom) that corresponds to the quantity $n - 1$. The t-value you see at that intersection is symbolized as $t_{1-P,n-1}$. Calculate your lower confidence limit as $\bar{x} - (SE_{\bar{x}}t_{1-P,n-1})$ and your upper confidence limit as $\bar{x} + (SE_{\bar{x}}t_{1-P,n-1})$. The entire 90 percent confidence interval is now presented formally as

$$P\{[\bar{x} - (SE_{\bar{x}}t_{.10,n-1})] \leq \mu \leq [\bar{x} + (SE_{\bar{x}}t_{.10,n-1})]\} = 0.90$$

The 95 percent confidence interval is presented as

$$P\{[\bar{x} - (SE_{\bar{x}}t_{.05,n-1})] \leq \mu \leq [\bar{x} + (SE_{\bar{x}}t_{.05,n-1})]\} = 0.95$$

The 99 percent confidence interval is presented as

$$P\{[\bar{x} - (SE_{\bar{x}}t_{.01,n-1})] \leq \mu \leq [\bar{x} + (SE_{\bar{x}}t_{.01,n-1})]\} = 0.99$$

For example, refer to design 10 in chapter 4. As an aside to the overall inquiry, for a public report that she must write, your supervisor asks that you provide her with an estimate of the mean number of forest bird species per UL forest tract, among all possible UL tracts in the cloud forest habitat. You decide to base the estimate on the sample of six replicate UL sites you've just examined there. You've calculated the sample statistics as $\bar{x} = 73$ species and $s^2 = 28$. Therefore, by equation 5.8 $SE_{\bar{x}}$ is 2.2. Given that $n = 6$, in table A.1 you look up t-values in the row for $df = 5$, in each of the three columns headed by $\alpha = 0.10$, $\alpha = 0.05$, and $\alpha = 0.01$, for the 90 percent, 95 percent, and 99 percent confidence levels, respectively. The t-value at those three intersections is 2.015, 2.571, and 4.032, respectively. Thus, the three confidence intervals are

$$P\{[73 - 2.2(2.015)] \leq \mu \leq [73 + 2.2(2.015)]\} = P\{69 \leq \mu \leq 77\} = 0.90$$

$$P\{[73 - 2.2(2.571)] \leq \mu \leq [73 + 2.2(2.571)]\} = P\{67 \leq \mu \leq 79\} = 0.95$$

$$P\{[73 - 2.2(4.032)] \leq \mu \leq [73 + 2.2(4.032)]\} = P\{64 \leq \mu \leq 82\} = 0.99$$

The working interpretation is, for example, that the interval of 67 to 79 is about 95 percent certain of including the true mean for the number of forest bird species that would be counted in each possible UL tract of cloud forest in the reserve. If this particular compromise between precision and confidence isn't satisfactory, your choices are to increase precision and sacrifice confidence (for example, the 90 percent interval) or to increase confidence but sacrifice precision (for example, the 99 percent interval). Of course, you could calculate confidence limits for any value of P, most easily for a P where the value of $(1 - P)$ is at the head of one of the columns in table A.1. There's nothing magic about the traditional compromise of selecting $P = 0.95$.

Procedures for calculating confidence limits for other parameters or population statistics (Sokal and Rohlf 1995) are analogous but not identical.

Table A.1. Values of the test statistic t. If calculating confidence limits for the population mean μ, use the value at the intersection of the column corresponding to $\alpha = (1 - P)$ and the row for df (degrees of freedom) $n - 1$. If estimating minimum sample size, follow the procedure specified in the appropriate section of appendix B. This table is reproduced with slight modification, with permission granted by Pearson Education, from table III in Fisher and Yates (1963).

df	α 0.9	0.8	0.7	0.6	0.5	0.4	0.3	0.2	0.1	0.05	0.02	0.01	0.001
1	0.158	0.325	0.510	0.727	1.000	1.376	1.963	3.078	6.314	12.706	31.821	63.657	636.62
2	0.142	0.289	0.445	0.617	0.816	1.061	1.386	1.886	2.920	4.303	6.965	9.925	31.598
3	0.137	0.277	0.424	0.584	0.765	0.978	1.250	1.638	2.353	3.182	4.541	5.841	12.924
4	0.134	0.271	0.414	0.569	0.741	0.941	1.190	1.533	2.132	2.776	3.747	4.604	8.610
5	0.132	0.267	0.408	0.559	0.727	0.920	1.156	1.476	2.015	2.571	3.365	4.032	6.869
6	0.131	0.265	0.404	0.553	0.718	0.906	1.134	1.440	1.943	2.447	3.143	3.707	5.959
7	0.130	0.263	0.402	0.549	0.711	0.896	1.119	1.415	1.895	2.365	2.998	3.499	5.408
8	0.130	0.262	0.399	0.546	0.706	0.889	1.108	1.397	1.860	2.306	2.896	3.355	5.041
9	0.129	0.261	0.398	0.543	0.703	0.883	1.100	1.383	1.833	2.262	2.821	3.250	4.781
10	0.129	0.260	0.397	0.542	0.700	0.879	1.093	1.372	1.812	2.228	2.764	3.169	4.587
11	0.129	0.260	0.396	0.540	0.697	0.876	1.088	1.363	1.796	2.201	2.718	3.106	4.437
12	0.128	0.259	0.395	0.539	0.695	0.873	1.083	1.356	1.782	2.179	2.681	3.055	4.318
13	0.128	0.259	0.394	0.538	0.694	0.870	1.079	1.350	1.771	2.160	2.650	3.012	4.221
14	0.128	0.258	0.393	0.537	0.692	0.868	1.076	1.345	1.761	2.145	2.624	2.977	4.140
15	0.128	0.258	0.393	0.536	0.691	0.866	1.074	1.341	1.753	2.131	2.602	2.947	4.073
16	0.128	0.258	0.392	0.535	0.690	0.865	1.071	1.337	1.746	2.120	2.583	2.921	4.015
17	0.128	0.257	0.392	0.534	0.689	0.863	1.069	1.333	1.740	2.110	2.567	2.898	3.965
18	0.127	0.257	0.392	0.534	0.688	0.862	1.067	1.330	1.734	2.101	2.552	2.878	3.922
19	0.127	0.257	0.391	0.533	0.688	0.861	1.066	1.328	1.729	2.093	2.539	2.861	3.883
20	0.127	0.257	0.391	0.533	0.687	0.860	1.064	1.325	1.725	2.086	2.528	2.845	3.850
21	0.127	0.257	0.391	0.532	0.686	0.859	1.063	1.323	1.721	2.080	2.518	2.831	3.819
22	0.127	0.256	0.390	0.532	0.686	0.858	1.061	1.321	1.717	2.074	2.508	2.819	3.792
23	0.127	0.256	0.390	0.532	0.685	0.858	1.060	1.319	1.714	2.069	2.500	2.807	3.767
24	0.127	0.256	0.390	0.531	0.685	0.857	1.059	1.318	1.711	2.064	2.492	2.797	3.745
25	0.127	0.256	0.390	0.531	0.684	0.856	1.058	1.316	1.708	2.060	2.485	2.787	3.725
26	0.127	0.256	0.390	0.531	0.684	0.856	1.058	1.315	1.706	2.056	2.479	2.779	3.707
27	0.127	0.256	0.389	0.531	0.684	0.855	1.057	1.314	1.703	2.052	2.473	2.771	3.690
28	0.127	0.256	0.389	0.530	0.683	0.855	1.056	1.313	1.701	2.048	2.467	2.763	3.674
29	0.127	0.256	0.389	0.530	0.683	0.854	1.055	1.311	1.699	2.045	2.462	2.756	3.659
30	0.127	0.256	0.389	0.530	0.683	0.854	1.055	1.310	1.697	2.042	2.457	2.750	3.646
40	0.126	0.255	0.388	0.529	0.681	0.851	1.050	1.303	1.684	2.021	2.423	2.704	3.551
60	0.126	0.254	0.387	0.527	0.679	0.848	1.046	1.296	1.671	2.000	2.390	2.660	3.460
120	0.126	0.254	0.386	0.526	0.677	0.845	1.041	1.289	1.658	1.980	2.358	2.617	3.373
∞	0.126	0.253	0.385	0.524	0.674	0.842	1.036	1.282	1.645	1.960	2.326	2.576	3.291

Deciding on Sample Size

Whether you're asking a noncomparative question or a comparative one, the first step is to obtain a set of *preliminary data* that can provide crude estimates of μ and σ^2 for the statistical population of concern. In the case of a comparative question, these data should come from the "control" level or some other baseline level of the design factor. Using these two values (usually labeled $\bar{x}_{est}$, s^2_{est}), you can estimate crudely the minimum sample size, or the minimum number of replicates per level in the case of a comparative question with discrete levels of the design factor, that you'll need in the upcoming study in order to reach the desired level of confidence in your inferences.

Where can you obtain values for $\bar{x}_{est}$ and s^2_{est}? The best way is to perform some preliminary sampling under the same conditions as those under which you'll conduct the study proper. If that isn't possible, borrow data from the most similar studies done elsewhere. If no such data are available, make the most educated guess possible. If you're dealing with morphological measurements of animals, such as lengths of adult male caimans, your educated guess might involve the quite crude "rule of thumb" that often $s \approx \bar{x}/10$ (see Sokal and Rohlf 1995). If you're dealing with an ecological response variable, such as counts of birds per evaluation unit, you might begin with the extraordinarily crude rule of thumb that often $s \approx \bar{x}$ (personal observation). Clearly, it's vastly preferable to have collected preliminary data yourself.

Sample Size for a Noncomparative Question

Let's say that you must estimate the number of kick-net samples to be taken per location in a stream in order to estimate adequately the population density of the nymphs of a certain mayfly species at that point (see chapter 8). To estimate the minimum sample size required for this or any other noncomparative study, two alternatives exist (see Merritt and Cummins 1984; Krebs 1989). Please note that in some cases you enter s^2_{est}, in other cases s_{est}, which is of course simply $\sqrt{s^2_{est}}$.

The "Relative Precision" Method

The goal here is to estimate the minimum number of samples that will lower the *standard error of the mean* ($SE_{\bar{x}}$) to some acceptable relative value (relative, that is, to the value of the mean). One choice might be to estimate the minimum number of samples needed to lower $SE_{\bar{x}}$ to 5 percent of the sample mean (Krebs 1989), although, as you'll see, this might be asking for unreasonable rigor if the observations vary a lot. Given that

$$SE_{\bar{x}} = \sqrt{s^2/n} \tag{B.1}$$

in this case we substitute ($0.05\bar{x}$) for ($SE_{\bar{x}}$):

$$0.05_{\bar{x}} = \sqrt{s^2/n} \tag{B.2}$$

and then solve the equation for n, the minimum sample size. That is, in this case

$$n \geq [s^2_{est}/0.0025\bar{x}^2_{est}] \tag{B.3}$$

Practice

1. Using equation B.3, calculate the number of repetitions required to sample adequately, with a relative error of 5 percent as defined above, the population density of the mayfly nymphs mentioned above where $\bar{x}_{est} = 17$ individuals per evaluation unit (kick-net sample) and $s^2_{est} = 37$.

2. You decide that you cannot possibly afford the time to take so many samples, so you decide to relax the "relative error" to 25 percent instead of 5 percent. What changes must be made in the equations above? What's n now? Note that in real life this might be a case where you'd prefer to convert the raw data to logarithms before beginning any calculations (see chapter 5, note 4).

The "Absolute Precision" Method

Here, the goal is to calculate the sample size that will provide your estimate of the population mean μ with the desired degree of precision (width of confidence interval) and confidence ($P = 1 - \alpha$). The procedure is a bit more complex than the previous one:

1. Choose a value for q, where $2q$ is the acceptable width of the confidence interval. That is, $q =$ the acceptable level of "fuzziness" of your estimate of μ, half the magnitude of the distance between the two confidence limits. The choice of q should be based on biological, logical, management, or pragmatic arguments.

2. Choose a value for P, the desired level of confidence. As usual, P could be, but does not have to be, 95 percent, or 0.95.

3. Consult table B.1 of this appendix and, without worrying about the definition of the statistic z (for purposes of this book, at least), find the value of z that corresponds to $\alpha = 1 - P$, i.e., $z_\alpha = z_{1-p}$.

4. Solve the crude and provisional equation

$$n_{prov} \geq (z_\alpha s_{est}/q)^2 \tag{B.4}$$

5. Now turn to table A.1, the table of t-values used in appendix A. Using the provisional value for n that equation B.4 produced in the previous step, solve the final equation:

$$n \geq (t_{\alpha,n_{prov}} \, s_{est}/q)^2 \qquad (B.5)$$

where $t_{\alpha,n_{prov}}$ is the tabulated value at the intersection between the column corresponding to α and the row corresponding to n degrees of freedom.

6. If the two consecutive values for n (the first, provisional value resulting from equation B.4 and the second value resulting from equation B.5) are identical once rounded off, you're done. If they aren't identical, though, you must return to table A.1 to obtain a new value for t—now that which corresponds to the n that resulted from the first application of equation B.5. Substitute the new t-value for the old one in equation B.5, redo the calculation, and if necessary continue with this *iterative* (circular) process until the value of n that you use to enter the equation (by employing its corresponding t-value) and the value of n with which you exit are the same after rounding. See table B.1.

Table B.1. Values of z corresponding to different values for α (or, in the procedure for comparative questions, for 2β).

α	z
0.001	3.30
0.002	3.10
0.0025	3.03
0.005	2.81
0.01	2.58
0.02	2.33
0.025	2.25
0.05	1.96
0.10	1.65
0.15	1.44
0.20	1.28
0.25	1.16
0.30	1.04
0.35	0.93
0.40	0.84
0.45	0.76
0.50	0.67
0.60	0.52
0.70	0.39
0.80	0.25

Example

Preliminary data on body mass of caimans are $\bar{x}_{est} = 29.6$ kg, $s_{est} = 2.7$, and you wish to estimate the minimum sample size that will be required, in the study proper, to provide an estimate of μ having 95 percent confidence limits of ± 1.2 kg (i.e., a 95 percent confidence interval 2.4 kg wide). The first step is to apply equation B.4. Given that table B.1 tells you that $z_{1-p} = z_{.05} = 1.96$,

$$n_{prov} \geq (1.96 \cdot 2.7/1.2)^2 = 19.4 \text{ caimans (round off to 19)}$$

Now you turn to table A.1 and find that $t_{.05,19} = 2.093$. Applying equation B.5:

$$n \geq (2.093 \cdot 2.7/1.2)^2 = 22.2 \text{ caimans (round off to 22)}$$

Uh-oh. You're not done, because the two values for n, 19 and 22, aren't the same. Therefore, you must follow an "iteration" process. Start over, but this time using 22 instead of 19. That is, turn again to table A.1 and look up the tabulated value for $t_{.05,22}$, which is 2.074. Now solve equation B.5 again, but this time using the new value. The result is

$$n \geq (2.074 \cdot 2.7/1.2)^2 = 21.8 \text{ caimans (round off to 22)}$$

Fine, *now* you're done: 21.8 rounds off to the value of 22, which is the same value you just used for obtaining the t-value for the second iteration of the equation. The iteration process is over. What can you conclude? In the study that follows this preliminary step, if and only if it's performed *under precisely the same conditions as those that produced the preliminary sample,* a sample of 22 caimans should yield a sample mean that, with 95 percent probability, will be within ±1.2 kg of μ.

Practice

Preliminary data on body mass of adult males in a population of endangered foxes yield the following: $\bar{x}_{est} = 4.7$ kg, $s_{est} = 1.2$. You wish to determine the minimum sample size that's required to yield, during the study proper, a sample mean that's within ±0.5 kg of μ with 99 percent confidence. What should the minimum n be? How does the minimum n change if you lower confidence to 95 percent? To 90 percent? How does the minimum n change if you're willing to accept a fuzzier estimate of ±1.0 kg? What should the minimum n be if you intend to achieve a precise estimate of ±0.25 kg?

Sample Size per Level for a Comparative Question

To estimate the minimum number of replicate response units required per level in a study with two categorical levels of the design factor, here's a tried and true recipe:

1. Decide on the acceptable risk of committing a Type I error should you decide to reject H_0, or α. As always, the traditional value for $\alpha_{rejection}$ is 0.05, but you might have a good reason for choosing 0.001, 0.01, 0.10, 0.20, 0.30, or some other value.

2. Decide on the acceptable risk of committing a Type II error should you decide not to reject H_0, or β. You might, for example, choose a β of 0.30, or 0.20, or 0.10 (that is, a power of 70 percent, 80 percent, or 90 percent, respectively).

3. Choose δ, the difference between levels in the mean value of the response variable that you wish to be able to detect with power = $1 - \beta$.

4. Using table B.1, solve the following crude and preliminary equation for n_{prov}:

$$n_{prov} \geq [2(z_\alpha + z_{2\beta})^2 s^2_{est}] / \delta^2 \tag{B.6}$$

   where n_{prov} is the first approximation of the minimum number of response units required per level of the design factor; s^2_{est} is, once again, the estimate that you're using for the variance; δ is the absolute value of the difference between μs that you wish to be able to detect; z_α is the tabulated value, in the second column of table B.1, that corresponds to the chosen α in the first column; $z_{2\beta}$ is the tabulated value in the second column of table B.1 that corresponds to the value in the first column that's equal to 2β.

5. Are you there yet? No. Turn to table A.1 to solve the "real" equation, beginning with the n_{prov} you've just obtained:

$$n \geq [2(t_{\alpha,n_{prov}-1} + t_{2\beta,n_{prov}-1})^2 s^2_{est}] / \delta^2 \tag{B.7}$$

   where $t_{\alpha,n_{prov}-1}$ is the tabulated value for the column whose heading corresponds to the chosen α and the row corresponding to $n_{prov} - 1$ degrees of freedom, and $t_{2\beta,n_{prov}-1}$ is the tabulated value at the intersection between the column whose heading corresponds to 2β and the row that corresponds to $n_{prov} - 1$ degrees of freedom.

6. If the two rounded-off values for n (that resulting from equation B.6 and that from equation

B.7) are identical, you're done. If the two values aren't equal, though, you must reenter equation B.7 using new values of t where n is now the number that just resulted from the first trial with equation B.7. As in the iterative process described previously, you must continue with iterations until the value for n entering the equation is the same as that exiting. This may take two or more iterations.

What have you done? Assuming that the preliminary data reflect without bias the nature of the statistical populations that you'll sample during the inquiry proper, you've just estimated the minimum number of replicate response units per level, n, needed to provide you with the power of $(1 - \beta)$ to detect an unknown true difference δ between the population means that correspond to the different levels of the design factor, with probability α of committing a Type I error should you reject the null hypothesis. Are you totally confused? If not, you're now armed with a powerful tool for completing the management cycle of figure 2.4. If so, please go back to the beginning of chapter 4 and start over.

Practice

You're concerned that adult females of an endangered lizard, in a site contaminated with mercury from gold-mining operations, may be stunted compared with lizards in a nearby uncontaminated site. The response variable is body mass. Note that you're comparing the two sites, one of which happens to have been contaminated with mercury and the other not; this design is clearly inadequate to address the effects of mercury contamination as the design factor. From preliminary data taken from lizards in a different uncontaminated site, you calculate $\bar{x}_{est} = 53.2$ g, $s^2 = 29$ (Note that you really don't need $\bar{x}_{est}$ for the calculations here). You want a design sufficiently strong to be able to detect an underlying δ of 5 g—the effect that you consider to be biologically significant based on your knowledge of the natural history of the lizards—between the μs of the two statistical populations corresponding to the two levels (sites), with $\alpha = 0.05$ and $\beta = 0.20$. About how many lizards, at the minimum, will you need to capture and weigh per site?

Repeat the exercise, but with each of the following variations in turn: (a) $\beta = 0.30$ (you're willing to run a greater risk of committing a Type II error); (b) $\beta = 0.05$ (you wish to lower the risk of committing a Type II error); (c) $\alpha = 0.01$ (you wish to exercise the utmost caution before deciding to reject the null hypothesis); (d) $\alpha = 0.10$ (you're willing to run a greater risk of committing a Type I error); (e) $\delta = 10$ g (you believe that a change of such magnitude is significant biologically, whereas a mean change of 5 g is not).

If you have the time and inclination, repeat for a broader range of values for α, β, and δ. How does the minimum required sample size change as you vary each one?

Important Notes

Note how the validity of the entire process, whether for a comparative or noncomparative question, depends on obtaining a reasonable $\bar{x}_{st}$ and s^2_{est}. Recall that you're estimating the *minimum* sample size that's required. In the noncomparative case, an n larger than that minimum would only increase the quality of your estimate. In the comparative case, larger sample sizes than the minimum would only increase the power of the statistical test (although very large sample sizes might also lead you to detect, with statistical significance, a biologically trivial δ). So it's always best to err on the side of caution—for example, by using a liberal value for s^2_{est}.

If you have a fixed sample size n and wish to detect a given δ while maintaining α at a reasonable value, such as 0.05, you can solve equations B.6 and B.7 for β in order to determine the predicted power of your statistical test to detect that δ under those particular conditions. *If these equations now produce negative values for* z *and* t, *or a nonsensical value for β, this signifies that β is precisely 1.0 and that the power of your test is 0*. That is, if a difference of size δ really did exist between the means of the statistical populations, with samples of size n you'd be unable to detect that with the α_{rej} you've chosen. This scenario may occur more often than you'd like to think. In such a case, it's possible that you can achieve a β that's less than 1.0 by allowing α_{rej} to increase, or by making the extra effort to increase n if that's possible, or by agreeing to choose a larger value for the δ that you wish to detect. Nevertheless, if the values of the response variable vary widely among themselves, it's also quite possible that none of these tactics will work unless you go to ridiculous extremes, such as an n of several thousand.

Note that you can easily perform all the computations with an inexpensive hand calculator or even with pencil and paper alone, as long as you have tables A.1 and B.1 with you. At the other extreme, you can choose from an increasingly large number of choices for computer software specifically written for power analysis. These software programs are not always correct, though, and they'll suggest procedures to you that are often inappropriate—such as *a posteriori* power analyses (Steidl, Hayes, and Schauber 1997; Gerard, Smith, and Weerakkody 1998; Johnson 1999).

As usual, I've presented only the very special case of deciding on the minimum n for a simple study design whose design factor has categorical levels, in this case just two. This procedure is adapted from those presented in Steel and Torrie (1960) and Sokal and Rohlf (1995); Krebs (1989) and Zar (1999) provide somewhat different but parallel approaches. Please note, though, that procedures exist to enable you to decide on sample sizes for many other designs or types of data, beginning with designs with three or more categorical levels of the design factor or designs with continuous levels. Techniques also exist for data that are ordinal or nominal instead of interval (see chapter 5). Borenstein and Cohen (1988) and Cohen (1988) are the most complete and accurate references available. Many people find them cumbersome to use because every design or statistical test requires a different procedure. Nevertheless, attempts to come up with a simpler, universal methodology (Kraemer and Thiemann 1987; Murphy and Myors 1998) have not been entirely satisfactory, even apart from the unfortunate fact that the second reference emphasizes *a posteriori* power analysis, which is not legitimate (Steidl, Hayes, and Schauber 1997; Gerard, Smith, and Weerakkody 1998). If you cannot obtain (or afford!) a copy of Cohen or Borenstein and Cohen, turn to Zar, who presents these procedures for several common study designs and statistical tests.

Appendix C

Resources Especially for Latin American Readers

Bibliographic Resources

The Essentials

As the text and the notes imply, you should buy, beg, or borrow the following without delay. They're listed in no special order.

- Margoluis and Salafsky (1998).

- Meffe and Carroll (1997).

- Sutherland (1996).

- At least one of the texts listed in note 12, chapter 8. If you're a Spanish speaker, for example, obtain one or both of the texts by Roldán Pérez (1992, 1996); if you're an English speaker, Rosenberg and Resh (1993) and/or Merritt and Cummins (1984).

- Siegel and Castellan in English (1988) or Spanish (1995).

- Any one of the statistics books mentioned in note 1, chapter 5: Steel and Torrie (1980, 1988), Sokal and Rohlf (1995) plus Rohlf and Sokal (1995), or Zar (1999).

- A subscription to the journal *Conservation Biology,* including back issues if possible. The journal is published by the Society for Conservation Biology. For information on membership in the society and subscriptions, send an email message to conbio@u.washington.edu, or write to Blackwell

Science, Inc., Society for Conservation Biology, Commerce Place, 350 Main Street, Malden, MA 02148-5018, USA.

- A subscription to the journal *Ecological Applications,* published by the Ecological Society of America. The society also publishes two key journals in basic ecology, *Ecology* and *Ecological Monographs.* Although fewer papers relevant to Latin American conservation concerns appear here than in *Conservation Biology,* those few may be very useful. Students and professionals in Latin America can obtain memberships in the society at reduced rates, with the option of receiving all three journals at a rock-bottom price. Write to Ecological Society of America, Member and Subscriber Services, 2010 Massachusetts Avenue, Washington, DC 20036, USA, or send an email message to esahq@esa.org.

Nice to Have Around

If feasible you should obtain access to, if not obtain outright, the following. Because of space limitations I won't list subscription information for the journals.

- *Journal of Wildlife Management.*

- *BioScience.*

- *Biological Conservation.*

- *Biotropica.*

- *Annual Review of Ecology and Systematics.*

- Krebs (1989).

- Southwood (1978).

- Laurance and Bierregaard (1997).

- Schelhas and Greenberg (1996).

- Primack et al. (2001).

- Forman (1997).

- A sampling from the numerous books spewing forth on conservation, management, sustainable development, and rural peoples in the latitudes where you live; these books, many of which are very good, appear so rapidly that if I were to list those current at this moment, they'd be superseded by the time you read this.

Of course, these two lists are far from being all-inclusive. For example, articles relevant to your conservation concerns or research ideas could appear in any one of several hundred journals. As the chapter notes suggest, numerous books exist on the themes developed here and related themes that may interest you. Nevertheless, with just the essentials on the first list you'll have the bibliographic support you need to carry out complete, objective inquiries.

Technical Resources

The Essentials

To be a conservation professional or field ecologist, you must have the following:

- An inexpensive battery- or sunlight-operated hand calculator that's reasonably weatherproof. Casio calculators tend to be more robust than those of any other brand. Seek a model that calculates simple sample statistics; you can easily buy two for well under US$40.

- A decent typewriter with a new ribbon. Inadequate typewriters with ribbons so old the ink is nearly invisible still infest the offices of many protected areas, government ministries, nongovernmental organizations, and academic or research institutions.

Nice to Have Around

For obvious reasons the following can be useful, but they're neither sufficient nor necessary to qualify you as a conservation professional or field ecologist:

- A reasonably modern personal computer (desktop or laptop) that is compatible with other reasonably modern computers. Remember that this is just a tool, not inherently any more useful than a typewriter or a hand calculator. The computer won't help you to think, judge, conclude, or decide.

- A computer printer with a new ribbon (if dot matrix) or a new cartridge (if laser or ink-jet).

- Access to email or at least a fax connection, for keeping in touch with human resources. Access to the Internet is nice but is less important.

Human Resources

For Help in General

If you're not already at one, universities and research institutes should be your first stop in the search for human resources. You may find statisticians, experts on various groups of animals and plants, experts on key ecological processes, applied biologists, and social scientists of all kinds. Don't think you must necessarily go to the most prestigious universities in the capital cities to find expert help. Excellent and expert colleagues often exist in smaller, provincial institutions. Natural history museums and herbaria, though, usually do reside in the capital city. Their staffs always include experts on various taxonomic groups but sometimes also employ experts in conceptual areas such as landscape ecology. The staffs of national or international research institutes in applied fields such as forestry or agriculture, sometimes but not always affiliated with universities, often employ large numbers of world-class professionals who can be of immense help to you. A very few examples include INTA in Argentina, CIAT in Colombia and elsewhere, CATIE in Costa Rica, Instituto de Ecología A.C. in Xalapa, Mexico, and INPA in Manaus, Brazil.

For Help with Ecological Indicators

If you need expert advice on the methodology for working with a particular target group, the suggestions above should give you a start. For example, by contacting local universities, museums, and herbaria, you'll almost certainly locate experts on commonly chosen target groups such as mammals, birds, frogs, snakes, butterflies, and plants. Amateurs may also be extraordinarily helpful; for example, in most countries you'll find formal or loosely knit groups dedicated to birds, orchids, or butterflies.

At first it may be more difficult to encounter local help with some of the less charismatic animals that make the best indicator groups, or with ecological processes as indicators, in Latin America and the Caribbean. Therefore, I'll list names of a few institutions and individuals willing to give you advice either directly or indirectly, by providing you with names of experts in your own region. Feel free to contact them, although please don't abuse their kindness. Also, please understand that the names listed represent a minute sample of the numerous highly qualified people in each case. For example, I've listed no experts for Brazil or the Caribbean nations simply because at the moment there's no one there, working in these particular areas, whom I know personally well enough to cajole into being on this list.

DUNG BEETLES AND RELATIVES

Dr. Gonzalo Halffter: halffter@ecologia.edu.mx

Dr. Mario Favila: favila@ecologia.edu.mx

Biól. Lucrecia Arellano: lucreci@ecologia.edu.mx

Biól. Federico Escobar: escobarf@ecologia.edu.mx

Instituto de Ecología
Km. 2.5 Carretera Antigua a Coatepec
Apartado Postal 63
91000 Xalapa, Veracruz, Mexico

EARTHWORMS

Dr. Carlos A. Fragoso
Instituto de Ecología (see above)
fragosoc@ecologia.edu.mx

BENTHIC INSECTS AND HOW TO DEVELOP A B-IBI

Dr. James R. Karr
P.O. Box 352020
222A Fishery Sciences
University of Washington
Seattle, WA 98195-5200, USA
jrkarr@u.washington.edu

Note also that excellent, helpful groups of faculty and students studying benthic insects as ecological indicators exist at many universities, for example in Latin America at Universidad del Valle and Universidad de Antioquia in Colombia, Universidad del Azuay in Ecuador, and Universidad Nacional de Tucumán (Instituto Miguel Lillo) in Argentina.

FRESHWATER FISHES AND HOW TO DEVELOP A FISH-IBI

Dr. James R. Karr (see above)

ECOLOGICAL PROCESSES AND INTERACTIONS

Fundación EcoAndina
A.A. 22527
Cali, Colombia

Dra. Carolina Murcia: cmurcia@uniweb.net.co

Dr. Gustavo Kattan: gukattan@cali.cetcol.net.co

Dr. Marcelo A. Aizen
Departamento de Ecología
Universidad Nacional del Comahue
Centro Regional Universitario Bariloche
8400 San Carlos de Bariloche, Argentina
marcito@cab.cnea.gov.ar or marcito@crub.uncoma.edu.ar

Dr. Peter Feinsinger
Department of Biological Sciences
Northern Arizona University
Flagstaff, AZ 86011, USA
peter.feinsinger@nau.edu

For Help with Extending the Reach of Inquiry

Chapter 10 presents a rather particular philosophy, which many colleagues and I have developed over the past decade or so. For information on schoolyard ecology in Latin America, for names of people in your region who are involved in this approach, and for information on obtaining the manual by Arango, Chaves, and Feinsinger (2001), contact:

Ricardo Stanoss
Latin American and Caribbean Program
National Audubon Society, U.S.A.
444 Brickell Avenue, Suite 850
Miami, FL 3313-1455, USA
rstanoss@audubon.org

Dr. Alejandro Grajal
Director, Latin American and Caribbean Program
National Audubon Society, U.S.A.
444 Brickell Avenue, Suite 850
Miami, FL 33131-1455, USA
agrajal@audubon.org

Or, if you have Internet access, see the Web pages:

http://www.audubon.org/local/latin/schoolyard.html
http://www.audubon.org/local/latin/EEPE

If you're interested in what's going on with schoolyard ecology in North America, contact the national coordinator of the SYEFEST program (School Yard Ecology for Elementary School Teachers):

Dr. Alan R. Berkowitz
Head of Education
Institute of Ecosystem Studies
P.O. Box R
Millbrook, NY 12545, USA
berkowitza@ecostudies.org
Web page: http://www.ecostudies.org

Please feel free to contact me directly (see address above) for names of people in your region of Latin America who are extending the inquiry approach to local communities, to park guards and other personnel of protected areas, or to visitors in protected areas.

Design and Statistics without the Jargon: A Play in Two Acts

You may be leading a workshop in "community inquiry." (See chapter 10.) Participants may have just voted on the question they most want answered. You're about to discuss ways to design the study itself. Or you may be working with park guards. The guards have just absorbed the somewhat different purposes of the management cycle (figure 2.4) and the field research cycle (figure 2.5). You may be about to focus on the all-important step of conceptual design. In either case, it's time to practice design and statistics.

The materials needed are: (a) a broom handle or stick that's at least 2 m long; (b) a roll of masking tape; and (c) indelible markers of two colors, preferably dark blue and either pink or red. The following, first design and then statistics, is presented as a stage play in two acts. If time is limited, you may need to stick with Act I alone. The script below provides only simple stage directions and your lines. The latter are quite stilted as written, but I'm assuming that you'll ad lib constantly. Clearly, you'll modify the vocabulary according to the nature of the group. The other participants will speak spontaneously, of course. Often they'll bring up many of the doubts, discoveries, and conclusions that I've included in your lines just in case. Keep the exercise interactive! If you take yourself too seriously, this could come across as being "politically incorrect." If you maintain your sense of humor throughout, though, everyone will have a great time and learn a great deal about design and statistics as well. Honest.

Act I

The participants are sitting down. You're in front pointing either at the arrow between "Question" and "Action" in figure 2.2 if this is a community workshop, or at the box entitled "Conceptual Design" in figures 2.4, 2.5 if it's a park guard workshop. You're speaking.

—Now, how are you going to get from the question to the answer? You can't just jump in and start taking data. First, you need to come up with a plan, or the design of your study. So, what's design? Design is the process of adjusting the scope of the sampling to the scope of the question, or, if you can't do that, the process of adjusting the scope of the question to the scope of your sampling. Why is design important? As an example I'll ask a very simple question and then design a study to answer it: "Who's taller, men or women?"

To answer that question I'll travel to the neighboring country of Lemuria. Lemuria has two large cities, Aguacate and Palta, and the people in these cities come from different ethnic backgrounds. On average, the Aguacatites are considerably taller than the Paltites. Now, all of you will become honorary citizens of Lemuria. Half of you will be men and half women, half Aguacatites and half Paltites.

[Ask the participants to line up from shortest to tallest. Assign the taller individuals to be Aguacatites and the shorter individuals to be Paltites, half and half. Among the Aguacatites, the taller individuals, regardless of their true gender, will be "men," indicated by bits of masking tape with a blue mark on their shoulders, and the shorter ones will be "women," indicated by bits of masking tape with a pink or red mark. Assign genders to the Paltites in the same manner. In each group, the Aguacatites and the Paltites, make one exception: make one woman taller than the shortest man. Finally, stick a long strip of clean masking tape along the broom handle or stick, all the way to the top.]

—OK, now I'm an absolutely brilliant scientist with the fascinating question "Who's taller, men or women?" Take note of the exact phrasing. Right now I'm in Aguacate. I'm going to answer the question by selecting a single individual of each gender and comparing their heights.

[Ask the Aguacatite "men" and "women" to form two separate groups by gender. Then with your eyes closed, randomly select one person from each group. Hold steady against the floor, vertically, the broom handle and use the colored marker to mark the height of each person on the masking tape, the man's height in blue and the woman's height in red. Display the broom handle to the whole group.]

—Wow! What a great scientific study I've done! See, the man is taller than the woman. This result fits with my preconceptions and with conventional wisdom, so I won't question it. I'll simply conclude that *men* are taller than *women*. Do you agree with my conclusion? Recall that I sampled those two people randomly. But by random sampling I might have chosen those other two people.

[Put new tape on the other side of the broom handle. Then with your eyes open, select the tallest "woman" and the shortest "man" among the Aguacatites and mark their heights on the tape as above.]

—Uh-oh. *The* man is shorter, not taller, than *the* woman. These results are contrary to those I'd expected and to conventional wisdom. Suddenly, I feel a great urge to review the methods I used and question them—there must be something wrong.

In either case, though, have I really addressed the question "Who's taller, *men* or *women*?" No? Then what is the question that I've truly addressed with this study design? Could it simply be "Who's taller, *this* man or *this* woman?" Is there any difference between the two questions? Can I say anything about the relative heights of *men* in the plural and *women* in the plural by measuring just one of each?

There are two very different points here. First, if a question specifies "men" plural and "women" plural, then we really must measure the heights of a representative sample of each gender. If we measure just one from each gender, then don't we have to change the question and the conclusions we draw? Second, if the results of an inquiry that's designed badly—or even one that's designed well—don't happen to match our preconceptions, then we tend to scrutinize the design and look for errors we made, or factors we didn't consider, that might explain the unexpected nature of the results. If the results fit with our expectations and with conventional wisdom, however, do we scrutinize the design and search for errors we made, or factors we didn't consider, just as carefully? Rarely. We unconsciously assume that all is fine. As in this case, though, couldn't a seriously flawed design lead to results that mimic those we expected just as easily as to results unlike those we expected? Moral: We should always scrutinize our studies with rigor no matter what the results may be (figure 2.2).

OK, as the Brilliant Scientist, now I realize my error. I still wish to answer the question "Who's taller, men or women?" Now I know that I must sample several of each. So now I'll go to Palta and do just that.

[Put fresh tape on the broom handle. Close your eyes and randomly select five men and five women from among the Paltites (fewer of each, if there aren't sufficient participants). On the tape, mark off their heights in blue and red, respectively. Estimate or eyeball the average height among the blue marks and that among the red marks. Indicate them on the tape.]

—Now my scientific brilliance astounds even me. Sure enough, in my sample men as a group are taller than women as a group. So, with a clear conscience I can conclude that men are taller than women. Do you agree with that conclusion?

But shouldn't my conclusion apply only to the scope of my sampling, Palta? Therefore, what question have I truly answered? A more limited one: "Who's taller in Palta, men or women?" I haven't answered the question for the whole world or even for Lemuria.

I do wish to extend my conclusions well beyond the city walls of Palta, though. So I'll expand the scope of my design so as to sample from both Aguacate and Palta. I conceive of a straightforward plan that will simplify my sampling and reduce the "noise" in my data: I'll sample men from Palta, women from Aguacate.

[Place fresh tape on the broomstick. Then with your eyes closed, randomly select five men from among the Paltites and five women from among the Aguacatites. Record their heights on the tape and estimate the two average heights by eye, as above.]

—Well now, that's a surprise! In my sample, the women's heights certainly are distinct from the men's heights, but what happened? What's wrong with my study design? Is there any other factor that might influence people's heights and that I failed to consider? What about the average difference in heights between Paltites as a group and Aguacatites as a group? So, was my design adequate to address the question "Who's taller in the Palta-Aguacate region, men or women?" No, you say? Then what's the question that my design truly addressed? How about, "Who's taller, men from Palta or women from Aguacate?" Isn't that rather a different question? Is it particularly interesting? What if instead I'd sampled women from Palta and men from Aguacate? In that case, once again the results would have conformed to my expectations. What question would the design really have addressed, though?

I'm now a sadder and humbler scientist. I realize that I must sample fairly from *both* Palta and Aguacate if I'm to extend my conclusions beyond the walls of just one city. So, I'll sample men and women from the Palta-Aguacate region regardless of their particular city of origin.

[Do just that. Put fresh tape on the broom handle. Ask all the women, Paltites and Aguacatites, to form a single group (warning them not to forget their city of origin) and all the men to form another group (ditto). Close your eyes and randomly select a fairly large sample of each gender. Open your eyes and tally the heights of all, as before. Almost certainly the red and blue marks will now be intermingled with one another. Eyeball the two average values and indicate them as always.]

—What a mess! On average it does look as if men in the Palta-Aguacate region are taller than women, but there's tremendous overlap and a lot of "noise," or variability, within each gender. The design was adequate for the question "Who's taller in the Palta-Aguacate zone, men or women?" but all that noise makes it difficult to get a clear answer. Is there an alternative design that might get rid of some of the noise? What if I sample equally from each of the two cities?

[Separate the Paltites and the Aguacatites again. Attach two fresh tapes to the broom handle, one to each side. Label them "A" and "P." Select at random five male Paltites, five female Paltites, five male Aguacatites, and five female Aguacatites. Using red and blue marks as always, record the heights of the Paltites on the "P" tape and the heights of the Aguacatites on the "A" tape. Eyeball them and indicate the mean height for each of the four groups.]

—Finally, I can see clearly what's going on. My data show that the pattern "men are taller on average than women" is consistent among the two cities. The average height varies considerably from Palta to Aguacate, but the *relative* difference between men and women is maintained from one to the other. This "block design," as it's called, enables me to distinguish the difference in heights between men and women from the difference in heights associated with the different cities. Here, the two cities are two different "blocks." This design isn't just adequate, it's a very strong design for answering my question—within limits. What was that question again—"Who's taller, men or women?" Have I truly answered that, or not? What, now you're telling me that I've answered only the much more limited question of "Who's taller *in the Palta-Aguacate zone,* men or women?" I still can't generalize from these results, produced by a beautifully designed study that encompasses two different cities, to the world at large or even just to Lemuria as a whole? Why not? What's the danger?

To illustrate the danger, let's say that I continue with the block design and sample men and women from several more Lemurian cities. The average height of the inhabitants continues to vary from city to city, but the relative difference between men's and women's heights remains consistent time after time. I become increasingly confident of my conclusion that men are taller, on average, than women throughout Lemuria. Finally, I sample from the city of Chirimoya—where I encounter the same pattern as always—and then visit the wealthy suburb of Guanábana just a short distance away.

[Put a fresh piece of masking tape on the broom handle.]

—Now every one of you is a Guanabanite, not a Paltite or Aguacatite.

[Again arrange everyone in the group by height, from shortest to tallest, regardless of true gender. Divide the line of people in the middle. All in the taller subgroup now get a red or pink marker, all in the shorter subgroup a blue marker—note the change.]

—As I enter Guanábana and look around, expecting one more confirmation of the pattern, I experience a rude shock. In Guanábana, all the men seem to be ugly, short, and fat—wealthy businessman, lawyers, and government officials. That's YOU!

[Point to the shorter, blue-marked individuals.]

—But all the women seem to be beautiful, elegant, slender, and tall: trophy wives. That's YOU!

[Point to the taller, red- or pink-marked individuals.]

—So, with a growing feeling of despair, I take my sample as always.

[Randomly select five of the "elegant ladies" and five of the "toads." Mark their individual heights on the new tape, in red and blue, respectively. Eyeball the two mean values and indicate these.]

—So much for my pattern. In answer to my original question, it seems that whether men are taller than women depends on just where you are.

This is the reason you should always exercise great caution in extending your conclusions too far beyond the scope of your sampling. There could always be a Guanábana just around the bend, a landscape or forest or community or population of the species you study where, for some reason, the patterns you'd come to expect no longer hold or actually reverse themselves. Whatever your inquiry and your intended use of the results, always be wary of Guanábanas not only in your design but also in your thinking afterward.

Now, what have we done? Let's review.

[Review the lesson learned at each point, beginning with the sample of one male and one female Aguacatite. If this is a park guard workshop, you might now turn to the selective logging question and the various designs discussed in chapter 4.]

Do people with no formal background in science, design, or statistics really absorb and retain all this? Yes. After all, strip it of the vocabulary (which is useful but not always necessary), and design is just common sense. I also suggest that you compare this exercise with the elements of chapter 4. Can you match any of the study designs meant to answer the height question with the designs, some better and some worse, for studies on the effects of selective logging? Next, turn to the glossary in box 4.1. In the height studies above, can you identify the design factor? The confounding factor, in at least one design? The levels of the design factor? Are the levels categorical or continuous? What's the response variable?

Act II

You're still measuring the heights of men and women, but now the questions are different and only one population is involved. The population has equal numbers of men and women. This time, you'll assign genders so that on average men are somewhat taller than women but individual heights overlap considerably: the heights of the three or four tallest women exceed the heights of the three or four shortest men.

—Now let's discuss a quite different perspective on how you might wish to generalize from a limited sample to a much bigger universe. The official name for this perspective is *statistical inference*. Like the philosophy of design, the philosophy of statistical inference is really just common sense. In this case, though, the philosophy has tremendous implications for decisions on biodiversity conservation, among other decisions.

All right, now you represent the population of all the people in the world. For just a moment, let's be omniscient and see what the true average height is across this huge number of people.

[Resuscitating the broom handle and replacing the masking tape, mark on it the height of every participant in the group, using the two colors for men and women as before. Eyeball the marks and indicate (a) the average height of women, (b) the average height of men, and (c) the average height across the entire group regardless of gender.]

—So this mark indicates the true average height among all people. In real life there's no way I could obtain that value, simply because it's impossible to measure the height of everyone in the world or even in a large town. If my question is simply "What's the average height of people?" all I can do is *estimate* that average height, basing my estimate on a much, much smaller random sample of people. I'll start with a sample of just one person.

[Place a fresh piece of tape on the broom handle opposite the tape that includes all the heights. Then with your eyes closed, randomly choose one person from the group. Record that person's height on the new tape.]

—With only one person, have I obtained an estimate of the average height among all people? Yes, I have. After all, this individual is one of those who contributed to that overall average—see, here's her [or his] height marked on the side from which that average came! With only one person I *do* have an estimate of the average. Of course, my estimate may not be very accurate. By chance it could be very distant from the true average value. On the other hand, by chance it could fall right on that average. In real life I have no way of knowing which. Being omniscient right now, though, we can check it out.

[Compare the height of the person sampled with the average value for the group as a whole, indicated on the previous tape.]

—Now, if I increase the number of people in my sample, will I increase the quality of my estimate? Let's see.

[If possible, put two more strips of tape on the broom handle at ninety degrees from the two already there and without hiding those. Select a random group of three people, then a random group of six. Record the heights of each group in red and blue on one of the two new tapes. Estimate the average value of each new sample by eye, without regard to gender. Compare these two average values with the "true" average as well as with the estimate based on the height of the single individual.]

—Sure enough, it appears that *on average* the larger my sample, the better my estimate, although by chance that might not hold in every case. Now, with a sample of any size could I ever be *absolutely sure* that I'd obtained the correct value of the average height for all people in the world? No? In theory, what's the only way I could be sure beyond the shadow of a doubt? By continuing until I'd measured every person in the world, right? That's impossible, though, so you're stuck with a sample that's much smaller.

What's the moral? No matter how large your sample, it's still just a sample from a much larger universe of possibilities. If you place absolute faith in the accuracy of your estimate, you always run the risk of being wrong.

Now let's change. Let's turn to a comparative question, whose answer will help me make an important decision. Instead of a scientist, now I'm the owner of a clothing factory, and I'm about to start manufacturing men's and women's blue jeans. I'm aware that on average men and women come in slightly different shapes, so I'll incorporate that fact into the manufacturing process. I don't know, though, whether I should also incorporate an average difference in height, because I'm not sure whether men on average really are taller than women. Whatever I decide will influence the machinery I'll purchase, the production process I'll set up, and perhaps my sales. If the decision is the wrong one, I'll suffer some economic consequences. So I need to answer the familiar question "On average, are men taller than women?" I'm wise enough to know that I must sample quite a few of each, but I'm also wise enough to know that I can't possibly measure the height of every single person in the world who might buy a pair of my blue jeans. So, as always, I'm going to sample from the population at large and base my decisions on the results of the sample. Let's pretend that I'm sampling randomly.

Refer back to the average heights of all men in the world, and of all women in the world, that you'd marked on the tape at the beginning of Act II. Now stick a fresh piece of tape on the opposite side of the broom handle from those two marks. Select a sample of five men and five women, but not randomly. This time, keep your eyes open and intentionally select individuals so that the average heights of the two quintets will be identical. On the fresh tape, mark the height of each of the ten people in the sample and then mark the eyeballed (identical) averages for the two genders.

—This contrived sample, which could just as easily have come about through genuinely random sampling of your heights, suggests that on average men and women have identical heights. So I use that result as the basis for a decision to invest in a single set of machinery and a single production process for men's and women's blue jeans, making only slight modifications for differences in body shape. Unbeknownst to me, though, by chance my sample wasn't representative. In the underlying population, the average heights of men and women truly are different. I have no way of knowing that, though, do I? So I've made what later turns out to be the wrong decision. And my wrong decision costs me a lot of money, because I've produced too many jeans for tall women and too many for short men.

Now, let's reverse the situation. Imagine that the average heights of men and women are the same in the underlying population. I ask my question as before and take my sample.

[On the tape that presents the heights of everyone in the room ("in the world"), circle the overall average height of all men and women taken together. State that now all women as one group, and all men as another group, have that same point as their average height; ignore the other marks on the tape. Place a fresh tape on the opposite side. With eyes open, intentionally select five men and five women so that the average height of the five men exceeds that of the five women. Eyeball those averages for each of the two samples and mark them on the fresh tape.]

—Here are the results from my sampling. There's a strong pattern apparent: in the sample, on average, men are taller than women. I use that as the basis for deciding to invest in two different sets of machinery, one for men's blue jeans and one for women's blue jeans, and I initiate a separate manufacturing process for each. Unbeknownst to me, though, by chance my sample wasn't representative. I've made what later turns out to be the wrong decision. I've invested money where I didn't need to. In the population at large, there's little height difference between men and women on average, so I could have invested in just one set of machinery.

What's the moral here? We're slaves to our samples. Whenever I use a sample to help me make a decision about a larger universe, I run the risk of being wrong. That is, if there's a pattern in the sample and I accept it as being true for the larger universe as well, I run the risk that I'm wrong and that such a pattern doesn't exist in the larger universe. If there's *no* apparent pattern in the sample and I accept that as being true for the larger universe, I run the risk that I'm wrong and that such a pattern does exist in the larger universe. In each case there may be a cost associated with being wrong. If I'm a blue jeans entrepreneur, my cost is merely financial.

Yes, I know, I know, that was a contrived and irrelevant example. Did you understand the point, though? If so, now let's switch to a subject of greater concern. Let's think about the ways we use the results of studies to guide us in making decisions on conservation issues or in proposing conservation guidelines. No matter how well we've designed the studies, no matter how many data we've collected, the results are just a sample of those that would be possible if we could extend the study over a larger landscape and for a longer time, right? All we have to work with is the sample. Either there's a pattern in the sample or there isn't.

[Modify the last statement as needed. For example, in a community workshop you might talk about decisions on crop management rather than on conservation guidelines. In any event, at this moment present a diagram of figure 5.4, modified as needed and without the terms "Type I error" and "Type II error," even though that's precisely what you're discussing.]

—If there's a strong apparent pattern in the results of a study we've just completed, if we conclude that there's something going on when unbeknownst to us there really isn't, and if we make decisions accordingly, what are the consequences to conservation [or to whatever is the pertinent theme]? If there's little apparent pattern in the results, if we then conclude that nothing is going on when unbeknownst to us there really *is*, and if we make decisions accordingly—or decide not to make them—what are the consequences? Which consequences might be more serious in the long run? And so which risk should we attempt most strenuously to minimize? How can we do that?

One part of the approach known as *statistical inference* enables us to judge how good our estimates might be—such as an estimate, based on a limited sample, of the mean height of people in the world at large. Another part, though, helps us to minimize the chances of making one or another of the two mistakes described above, or at least to estimate the *risk* of making those mistakes given the particular sample of data we've collected. Many specific techniques in this branch of statistical inference are labeled "statistical tests." Traditionally, most such tests help us to estimate the risk of being wrong if, based on our results, we conclude that there's a pattern in the larger universe that our sample represents. It's much more difficult, and much less traditional, to estimate the other risk, that of being wrong if we conclude that there's no pattern in the larger universe that our data represent. In any case, *neither statistical tests nor other mathematical techniques are of any help at all in evaluating the real-life consequences of making one mistake or the other*, whether the consequences are financial losses to the blue jeans entrepreneur or biodiversity losses to the conservation professional.

Unless you have some background in using statistical tests appropriately, don't worry about them. Just use your common sense. Whatever your inquiry, first be as sure as possible that your design is appropriate to your question, and vice versa. Then, try for as large a sample as is reasonable. Remember that whatever pattern or lack of pattern that shows up in your data could always have come about by chance alone. Recognize that whatever you conclude based on your results, whatever decisions you take after reflecting, you always run some risk of being wrong. Consider that being wrong can have consequences, some more serious than others.

That's it for the basic philosophy of statistical inference, especially the philosophy of Type I and Type II errors and their consequences to conservation. Again, will people with no formal background in science or design or statistics really absorb and apply all this? Yes, except perhaps for the last little digression on the meaning of statistical inference and statistical tests.

Literature Cited

Agosti, D., J. D. Majer, L. E. Alonso, and T. R. Schultz. 2000. *Ants: Standards for measuring and monitoring biodiversity.* Smithsonian Institution Press, Washington, DC, USA.

Agresti, A. 1990. *Categorical data analysis.* Wiley-Interscience, New York, NY, USA.

Aizen, M. A., and P. Feinsinger. 1994. Forest fragmentation, pollination, and plant reproduction in a Chaco dry forest, Argentina. *Ecology* 75: 330–351.

Allan, J. D., and A. S. Flecker. 1993. Biodiversity conservation in running waters. *BioScience* 43: 32–43.

Alpert, P. 1995. Applying ecological research at integrated conservation and development projects. *Ecological Applications* 5: 857–860.

Andersen, A. N. 1997. Using ants as bioindicators: Multiscale issues in ant community ecology. *Conservation Ecology* [online] 1(1): 8. Available from the Internet. URL: *http://www.consecol.org/vol1/iss1/art8.*

Andrade, G. I., and H. Rubio-Torgler. 1994. Sustainable use of the tropical rain forest: Evidence from the avifauna in a shifting-cultivation habitat mosaic in the Colombian Amazon. *Conservation Biology* 8: 545–554.

Arango, N., M. E. Chaves, and P. Feinsinger. 2001. *Ecología en el patio escolar: Una guía metodológica para América Latina.* Latin American Program, National Audubon Society, New York, NY, USA.

Arnqvist, G., and D. Wooster. 1995. Meta analysis: Synthesizing research findings in ecology and evolution. *Trends in Ecology and Evolution* 10: 236–240.

Bawa, K. S., and A. Markham. 1995. Climate change and tropical forests. *Trends in Ecology and Evolution* 10: 348–349.

Bawa, K. S., and R. Seidler. 1998. Natural forest management and conservation of biodiversity in tropical forests. *Conservation Biology* 12: 46–55.

Beier, P., and R. F. Noss. 1998. Do habitat corridors provide connectivity? *Conservation Biology* 12: 1241–1252.

Binford, M. W., A. L. Kolata, M. Brenner, J. W. Janusek, M. T. Seddon, M. Abbott, and J. H. Curtis. 1997. Climate variation and the rise and fall of an Andean civilization. *Quaternary Research* 47: 235–248.

Bongers, T., and H. Ferris. 1999. Nematode community structure as a bioindicator in environmental monitoring. *Trends in Ecology and Evolution* 14: 224–228.

Borenstein, M., and J. Cohen. 1988. *Statistical power analysis.* Lawrence Erlbaum Associates, Hillsdale, NJ, USA.

Bright, C. 1998. *Life out of bounds: Bioinvasion in a borderless world.* W. W. Norton, New York, NY, USA.

Brower, L. P. 1997. A new paradigm in biodiversity conservation: endangered biological phenomena. Pages 115–118 in G. K. Meffe and C. R. Carroll, *Principles of conservation biology,* 2nd edition. Sinauer Associates, Sunderland, MA, USA.

Brown, J. H., and A. Kodric-Brown. 1977. Turnover rates in insular biogeography: Effect of immigration on extinction. *Ecology* 58: 445–449.

Brown, J. H., C. G. Curtin, and R. W. Braithwaite. In press. Management of the semi-natural matrix. In G. A. Bradshaw, P. A. Marquet, and H. A. Mooney, editors, *How landscapes change: Human disturbance and ecosystem disruptions in the Americas.* Springer-Verlag, New York, NY, USA.

Bucher, E. H. 1987. Herbivory in arid and semi-arid regions of Argentina. *Revista Chilena de Historia Natural* 60: 265–273.

Buchmann, S. L., and G. P. Nabhan. 1996. *The forgotten pollinators.* Island Press, Washington, DC, USA.

Bush, M. B., and P. A. Colinvaux. 1994. Tropical forest disturbance: Paleoecological records from Darien, Panama. *Ecology* 75: 1761–1768.

Camus, P. A., and M. Lima. 1995. El uso de la experimentación en ecología: Fuentes de error y limitaciones. *Revista Chilena de Historia Natural* 68: 19–42.

Caro, T. M., and G. O'Doherty. 1999. On the use of surrogate species in conservation biology. *Conservation Biology* 13: 805–814.

Carrillo, E., G. Wong, and A. D. Cuáron. 2000. Monitoring animal populations in Costa Rican protected areas under different hunting restrictions. *Conservation Biology* 14: 1580–1591.

Castillo, A., and V. M. Toledo. 2000. Applying ecology in the third world: The case of Mexico. *BioScience* 50: 66–76.

Chapman, C. A., S. R. Balcomb, T. R. Gillespie, J. P. Skorupa, and T. T. Struhsaker. 2000. Long-term effects of logging on African primate communities: A 28-year comparison from Kibale National Park, Uganda. *Conservation Biology* 14: 207–217.

Coblentz, B. E. 1990. Exotic organisms: A dilemma for conservation biology. *Conservation Biology* 4: 261–265.

Cohen, J. 1988. *Statistical power analysis for the behavioral sciences,* 2nd edition. Lawrence Erlbaum Associates, Hillsdale, NJ, USA.

Colwell, R. K., and J. A. Coddington. 1994. Estimating terrestrial biodiversity through extrapolation. *Philosophical Transactions of the Royal Society of London,* Series B 345: 101–118.

Conover, W. J. 1980. *Practical nonparametric statistics,* 2nd edition. John Wiley, New York, NY, USA.

Cook, T. D., and D. T. Campbell. 1979. *Quasi-experimentation: Design and analysis issues for field settings.* Houghton Mifflin, Boston, MA, USA.

Cooperrider, A. Y. 1996. Science as a model for ecosystem management: Panacea or problem? *Ecological Applications* 6: 736–737.

Crome, F. H. J. 1997. Researching tropical forest fragmentation: Shall we keep on doing what we're doing? Pages 485–501 in W. F. Laurance and R. O. Bierregaard Jr., *Tropical forest remnants: Ecology, management, and conservation of fragmented communities.* University of Chicago Press, Chicago, IL, USA.

Crome, F. H. J., M. R. Thomas, and L. A. Moore. 1996. A novel Bayesian approach to assessing impacts of rain forest logging. *Ecological Applications* 6: 1104–1123.

Cronk, Q. C. B., and J. L. Fuller. 1995. Plant invaders: The threat to natural ecosystems. *Trends in Ecology and Evolution* 10: 508–509.

Crowley, P. H. 1992. Resampling methods for computation-intensive data analysis in ecology and evolution. *Annual Review of Ecology and Systematics* 23: 405–447.

Dafni, A. 1992. *Pollination ecology: A practical approach.* Oxford University Press, Oxford, UK.

Dale, V. H., S. Brown, R. A. Haeuber, N. T. Hobbs, N. T. Huntly, R. J. Naiman, W. E. Riebsame, M. G. Turner, and T. J. Valone. 2000. Ecological principles and guidelines for managing the use of land. *Ecological Applications* 10: 639–670.

Dale, V. H., and R. A. Haeuber, editors. 2000. Forum: Perspectives on land use. *Ecological Applications* 10: 671–688.

Damascos, M. A., and G. G. Gallopin. 1992. Ecología de un arbusto introducido (*Rosa rubiginosa* L. = *Rosa eglanteria* L.): Riesgo de invasión y efectos en las comunidades vegetales de la región andino-patagónica de Argentina. *Revista Chilena de Historia Natural* 65: 395–407.

Daniel, W. W. 1990. *Applied nonparametric statistics*. PWS-KENT, Boston, MA, USA.

Darwin, C. 1881. *The formation of vegetable mould through the action of worms: With observations on their habits*. J. Murray, London, UK.

Davis, M. B., editor. 1990. Special issue: Biology and palaeobiology of global climate change. *Trends in Ecology and Evolution* 5: 269–322.

de Lima, M. G., and C. Gascon. 1999. The conservation value of linear forest remnants in central Amazonia. *Biological Conservation* 91: 241–247.

DeLuca, T. H., W. A. Patterson IV, W. A. Freimund, and D. N. Cole. 1998. Influence of llamas, horses, and hikers on soil erosion from established recreation trails in western Montana, USA. *Environmental Management* 22: 255–262.

Denevan, W. M. 1992. The pristine myth: The landscape of the Americas in 1492. *Annals of the Association of American Geographers* 82: 369–385.

Denslow, J. S. 1987. Tropical rainforest gaps and tree species diversity. *Annual Review of Ecology and Systematics* 18: 431–452.

Devine, R. 1998. *Alien invasion: America's battle with non-native animals and plants*. National Geographic Society, Washington, DC, USA.

Dias, P. C. 1996. Sources and sinks in population biology. *Trends in Ecology and Evolution* 11: 326–330.

Didham, R. K., J. Ghazoul, N. E. Stork, and A. J. Davis. 1996. Insects in fragmented forests: A functional approach. *Trends in Ecology and Evolution* 11: 255–260.

Digby, P. G. N., and R. A. Kempton. 1987. *Multivariate analysis of ecological communities*. Chapman and Hall, London, UK.

Dirzo, R., and A. Miranda. 1991. Altered patterns of herbivory and diversity in the forest understory: A case study of the possible consequences of contemporary defaunation. Pages 273–287 in P. W. Price, T. M. Lewinsohn, G. W. Fernandes, and W. W. Benson, editors, *Plant-animal interactions: Evolutionary ecology in tropical and temperate regions*. John Wiley, New York, NY, USA.

Dixon, P., and A. M. Ellison, editors. 1996. Bayesian inference. *Ecological Applications* 6: 1034–1123.

Dutilleul, P. 1993. Spatial heterogeneity and the design of ecological field experiments. *Ecology* 74: 1646–1658.

Eberhardt, L. L., and J. M. Thomas. 1991. Designing environmental field studies. *Ecological Monographs* 61: 57–73.

Estades, C. F., and S. A. Temple. 1999. Deciduous-forest bird communities in a fragmented landscape dominated by exotic pine plantations. *Ecological Applications* 9: 573–585.

Farnsworth, E. J., and J. Rosovsky. 1993. The ethics of ecological field experimentation. *Conservation Biology* 7: 463–472.

Favila, M. E., and G. Halffter. 1997. The use of indicator groups for measuring biodiversity as related to community structure and function. *Acta Zoológica Mexicana* (n.s.) 72: 1–25.

Feinsinger, P. 1976. Organization of a tropical guild of nectarivorous birds. *Ecological Monographs* 46: 257–291.

———. 1997. Habitat "shredding." Pages 270–271 in G. K. Meffe and C. R. Carroll, *Principles of conservation biology*, 2nd edition. Sinauer Associates, Sunderland, MA, USA.

Feinsinger, P., and R. K. Colwell. 1978. Community organization among Neotropical nectar-feeding birds. *American Zoologist* 18: 779–795.

Feinsinger, P., K. G. Murray, S. Kinsman, and W. H. Busby. 1986. Floral neighborhood and pollination success in four hummingbird-pollinated Costa Rican plant species. *Ecology* 67: 449–464.

Feinsinger, P., J. H. Beach, Y. B. Linhart, W. H. Busby, and K. G. Murray. 1988. Mixed support for spatial heterogeneity in species interactions: Hummingbirds in a tropical disturbance mosaic. *American Naturalist* 131: 33–47.

Feinsinger, P., H. M. Tiebout III, and B. E. Young. 1991. Do tropical bird-pollinated plants exhibit density dependent interactions? Field experiments. *Ecology* 72: 1953–1963.

Feinsinger, P., L. Margutti, and R. D. Oviedo. 1997. School yards and nature trails: Ecology education outside the university. *Trends in Ecology and Evolution* 12: 115–120.

Fernández-Duque, E., and C. Valeggia. 1994. Meta-analysis: A valuable tool in conservation research. *Conservation Biology* 8: 555–561.

Fimbel, R., A. Grajal, and J. Robinson, editors. 2001. *Conserving wildlife in managed tropical forests.* Columbia University Press, New York, NY, USA.

Fisher, R. A., and F. Yates. 1963. *Statistical tables for biological, agricultural, and medical research.* Hafner Publishing Company, New York, NY, USA.

Flecker, A. S., and C. R. Townsend. 1994. Community-wide consequences of trout introduction in New Zealand streams. *Ecological Applications* 4: 798–807.

Fore, L. S., J. R. Karr, and R. W. Wisseman. 1996. Assessing invertebrate responses to human activities: Evaluating alternative approaches. *Journal of the North American Benthological Society* 15: 212–231.

Forman, R. T. T. 1997. *Land mosaics: The ecology of landscapes and regions.* Cambridge University Press, Cambridge, UK.

Forman, R. T. T., and M. Godron. 1986. *Landscape ecology.* John Wiley, New York, NY, USA.

Forman, R. T. T., and L. E. Alexander. 1998. Roads and their major ecological effects. *Annual Review of Ecology and Systematics* 29: 207–232.

Foster, D. R., P. K. Schoonmaker, and S. T. A. Pickett. 1990. Insights from paleoecology to community ecology. *Trends in Ecology and Evolution* 5: 119–122.

Fragoso, C. 1989. Las lombrices de tierra de la reserva "El Cielo": Aspectos ecológicos y sistemáticos. *BIOTAM* 1: 38–44.

———. 1992. Las lombrices terrestres de la Selva Lacandona: Sistemática, ecología y potencial práctico. Pages 101–118 in M. A. Vásquez-Sánchez and M. A. Ramos, editors, *Reserva de la Biósfera Montes Azules, Selva Lacandona: Investigación para su conservación.* Publicaciones Especiales Ecosfera 1. Centro de Estudios para la Conservación de Recursos Naturales, A.C. Mexico D.F., Mexico.

Fragoso, C., I. Barois, C. González, C. Arteaga, and J. C. Patrón. 1993. Relationship between earthworms and soil organic matter levels in natural and managed ecosystems in the Mexican tropics. Pages 231–239 in K. Mulongoy and R. Merckx, editors, *Soil organic matter dynamics and sustainability of tropical agriculture.* Wiley-Sayce, London, UK.

Fretwell, S. D. 1972. *Populations in a seasonal environment.* Princeton University Press, Princeton, NJ, USA.

Fritts, T. H., and G. H. Rodda. 1998. The role of introduced species in the degradation of island ecosystems: A case history of Guam. *Annual Review of Ecology and Systematics* 29: 113–140.

Frumhoff, P. C. 1995. Conserving wildlife in tropical forests managed for timber. *BioScience* 45: 456–464.

———. 1996. *Monitoring the environmental impacts of USAID-funded activities to conserve biological diversity.* U.S. Agency for International Development, ENRIC (Environment and Natural Resources Information Center), Washington, DC, USA.

Gaines, S. D., and M. W. Denny. 1993. The largest, smallest, highest, lowest, longest, and shortest: Extremes in ecology. *Ecology* 74: 1677–1692.

Gascon, C., T. E. Lovejoy, R. O. Bierregaard Jr., J. R. Malcolm, P. C. Stouffer, H. L. Vasconcelos, W. F. Laurance, B. Zimmerman, M. Tocher, and S. Borges. 1999. Matrix habitat and species richness in tropical forest remnants. *Biological Conservation* 91: 223–229.

Gauch, H. G., Jr. 1982. *Multivariate analysis in community ecology.* Cambridge University Press, Cambridge, UK.

Gerard, P. D., D. R. Smith, and G. Weerakkody. 1998. Limits of retrospective power analysis. *Journal of Wildlife Management* 62: 801–807.

Gerrodette, T. 1987. A power analysis for detecting trends. *Ecology* 68: 1364–1372.

Gibbons, A. 1990. New view of early Amazonia. *Science* 248: 1488–1490.

Gómez-Pompa, A. and A. Kaus. 1992. Taming the wilderness myth. *BioScience* 42: 271–279.

Gonick, L., and W. Smith. 1993. *The cartoon guide to statistics.* HarperCollins, New York, NY, USA.

González, G., X. Zou, and S. Borges. 1996. Earthworm abundance and species composition in abandoned tropical croplands: Comparisons of tree plantations and secondary forests. *Pedobiologia* 40: 385–391.

Gotelli, N. J., and G. R. Graves. 1996. *Null models in ecology.* Smithsonian Institution Press, Washington, DC, USA.

Graber, D. M. 1995. Resolute biocentrism: The dilemma of wilderness in national parks. Pages 123–135 in M. E. Soulé and G. Lease, editors, *Reinventing nature?* Island Press, Washington, DC, USA.

Graham, R. W. 1986. Response of mammalian communities to environmental changes during the late Quaternary. Pages 300–313 in J. M. Diamond, and T. J. Case, editors, *Community ecology.* Harper and Row, New York, NY, USA.

———. 1988. The role of climatic change in the design of biological reserves: The paleoecological perspective for conservation biology. *Conservation Biology* 2: 391–394.

Greig-Smith, P. 1983. *Quantitative plant ecology,* 3rd edition. University of California Press, Berkeley, CA, USA.

Griffith, D. M. 2000. Agroforestry: A refuge for tropical biodiversity after fire. *Conservation Biology* 14: 325–326.

Guariguata, M. R. 1990. Landslide disturbance and forest regeneration in the upper Luquillo Mountains of Puerto Rico. *Journal of Ecology* 78: 814–832.

Gurevitch, J., and L. V. Hedges. 1999. Statistical issues in ecological meta-analyses. *Ecology* 80: 1142–1149.

Hackel, J. D. 1999. Community conservation and the future of Africa's wildlife. *Conservation Biology* 13: 726–734.

Haemig, P. D. 1978. Aztec emperor Auitzotl and the great-tailed grackle. *Biotropica* 10: 11–17.

———. 1979. Secret of the painted jay. *Biotropica* 11: 81–87.

Halffter, G. 1998. A strategy for measuring landscape biodiversity. *Biology International* 36: 3–17.

Halffter, G., M. E. Favila, and V. Halffter. 1992. A comparative study of the structure of the scarab guild in Mexican tropical rain forests and derived ecosystems. *Folia Entomológica Mexicana* 82: 195–238.

Halffter, G., and M. E. Favila. 1993. The Scarabaeinae (Insecta: Coleoptera), an animal group for analysing, inventorying and monitoring biodiversity in tropical rainforest and modified landscapes. *Biology International* 27: 15–21.

Halpin, P. N. 1997. Global climate change and natural-area protection: Management responses and research directions. *Ecological Applications* 6: 828–843.

Hamburg, S. P., and R. L. Sanford Jr. 1986. Disturbance, *Homo sapiens,* and ecology. *Bulletin of the Ecological Society of America* 67: 169–170.

Hamilton, A., A. Cunningham, D. Byarugaba, and F. Kayanja. 2000. Conservation in a region of political instability: Bwindi Impenetrable Forest, Uganda. *Conservation Biology* 14: 1722–1725.

Hanski, I., and Y. Cambefort, editors. 1991. *Dung beetle ecology.* Princeton University Press, Princeton, NJ, USA.

Hargrove, W. W., and J. Pickering. 1992. Pseudoreplication: A *sine qua non* for regional ecology. *Landscape Ecology* 6: 251–258.

Hartshorn, G. S. 1995. Ecological basis for sustainable development in tropical forests. *Annual Review of Ecology and Systematics* 26: 155–176.

Haslett, J. R. 1990. Geographic information systems: A new approach to habitat definition and the study of distributions. *Trends in Ecology and Evolution* 5: 214–218.

Heinz, D. 1995. Continued grazing, a presupposition. *Conservation Biology* 9: 708–709.

Hendrix, P. D., editor. 1995. *Earthworm ecology and biogeography in North America*. Lewis Publishers, Boca Raton, FL, USA.

Herrera, C. M. 1998. Long-term dynamics of Mediterranean frugivorous birds and fleshy fruits: A 12-year study. *Ecological Monographs* 68: 511–538.

Heyer, W. R., M. A. Donnelly, R. W. McDiarmid, L.-A. C. Hayek, and M. S. Foster, editors. 1994. *Measuring and monitoring biological diversity: Standard methods for amphibians*. Smithsonian Institution Press, Washington, DC, USA.

Hill, K., J. Padwe, C. Bejyvagi, A. Bepurangi, F. Jakugi, R. Tykuarangi, and T. Tykuarangi. 1997. Impact of hunting on large vertebrates in the Mbaracayu Reserve, Paraguay. *Conservation Biology* 11: 1339–1353.

Hilty, J., and A. Merenlender. 2000. Faunal indicator taxa selection for monitoring ecosystem health. *Biological Conservation* 92: 185–197.

Hobbs, R. J., and L. F. Huenneke. 1992. Disturbance, diversity, and invasion: Implications for conservation. *Conservation Biology* 6: 324–337.

Holling, C. S. 1978. Adaptive environmental assessment and management. John Wiley, New York, NY, USA.

Hourdequin, M. 2000. Introduction to Special Section: Ecological effects of roads. *Conservation Biology* 14: 16–17.

Hughes, L. 2000. Biological consequences of global warming: Is the signal already apparent? *Trends in Ecology and Evolution* 15: 56–61.

Hunter, J. L., G. L. Jacobson Jr., and T. Webb III. 1988. Paleoecology and the coarse-filter approach to maintaining biological diversity. *Conservation Biology* 2: 375–385.

Hunter, M., Jr., 1996. Benchmarks for managing ecosystems: Are human activities natural? *Conservation Biology* 10: 695–697.

Hurlbert, S. H. 1971. The nonconcept of species diversity: A critique and alternative parameters. *Ecology* 52: 577–586.

———, editor. 1981. *Aquatic biota of tropical South America*, vol. I and II. San Diego State University Press, San Diego, CA, USA.

———. 1984. Pseudoreplication and the design of ecological field experiments. *Ecological Monographs* 54: 187–211.

James, F. C., and C. E. McCullough. 1990. Multivariate analysis in ecology and systematics: Panacea or Pandora's box? *Annual Review of Ecology and Systematics* 21: 129–166.

Janzen, D. H. 1983. No park is an island: Increase in interference from outside as park size decreases. *Oikos* 41: 402–410.

———. 1986a. Chihuahuan desert nopaleras: Defaunated big mammal vegetation. *Annual Review of Ecology and Systematics* 17: 595–630.

———. 1986b. Lost plants. *Oikos* 46: 129–131.

———. 1995. Neotropical restoration biology. *Vida Silvestre Neotropical* 4: 3–9.

Janzen, D. H., and P. S. Martin. 1982. Neotropical anachronisms: The fruits the gomphotheres ate. *Science* 215: 19–27.

Johnson, D. H. 1995. Statistical sirens: The allure of nonparametrics. *Ecology* 76: 1998–2000.

———. 1999. The insignificance of statistical hypothesis testing. *Journal of Wildlife Management* 63: 763–772.

Jongman, R. H., C. J. F. ter Braak, and O. F. R. van Tongeren, editors. 1995. *Data analysis in community and landscape ecology*. Cambridge University Press, Cambridge, UK.

Kareiva, P. M., J. G. Kingsolver, and R. B. Huey, editors. 1993. *Biotic interactions and global change*. Sinauer Associates, Sunderland, MA, USA.

Karr, J. R. 1991. Biological integrity: A long-neglected aspect of water resource management. *Ecological Applications* 1: 66–84.

———. 1992. Measuring biological integrity: Lessons from streams. Pages 83–104 in S. Woodley, J. Kay, and

G. Francis, editors. *Ecological integrity and the management of ecosystems.* St. Lucie Press, Delray Beach, FL, USA.

———. 1996. Ecological integrity and ecological health are not the same. Pages 97–109 in P. C. Schulze, editor, *Engineering within ecological constraints.* National Academy Press, Washington, DC, USA.

———. 1998. Biological integrity: A long-neglected aspect of environmental program evaluation. Pages 148–175 in G. J. Knapp and T. J. Kim, editors, *Environmental program evaluation: A primer.* University of Illinois Press, Urbana, IL, USA.

Karr, J. R., and E. W. Chu. 1997. *Biological monitoring and assessment: Using multimetric indices effectively,* EPA 235-R97-001. University of Washington, Seattle, WA, USA.

———. 1998. *Restoring life in running waters: Better biological monitoring.* Island Press, Washington, DC, USA.

Kearns, C. A., and D. W. Inouye. 1993. *Techniques for pollination biologists.* University Press of Colorado, Niwot, CO, USA.

Kearns, C. A., D. W. Inouye, and N. M. Waser. 1998. Endangered mutualisms: The conservation of plant-animal interactions. *Annual Review of Ecology and Systematics* 29: 83–112.

Kerans, B. L., and J. R. Karr. 1994. A benthic index of biotic integrity (B-IBI) for rivers of the Tennessee Valley. *Ecological Applications* 4: 768–785.

Khan, A., M. Polak, D. Raaimakers, and R. Zagt. 1995. Basic and applied research for sound rain forest management in Guyana. *Ecological Applications* 5: 904–910.

Kiester, A. R., J. M. Scott, B. Csuti, R. F. Noss, B. Butterfield, K. Sahr, and D. White. 1996. Conservation prioritization using GAP data. *Conservation Biology* 10: 1332–1342.

Klein, B. C. 1989. Effects of forest fragmentation on dung and carrion beetle communities in central Amazonia. *Ecology* 70: 1715–1725.

Kotliar, N. B. 2000. Application of the new keystone-species concept to prairie dogs: How well does it work? *Conservation Biology* 14: 1715–1721.

Kraemer, H. Ch., and S. Thiemann. 1987. *How many subjects? Statistical power analysis in research.* Sage Publications, Newbury Park, CA, USA.

Krebs, C. J. 1989. *Ecological methodology.* HarperCollins, New York, NY, USA.

Kremen, C. 1992. Assessing the indicator properties of species assemblages for natural areas monitoring. *Ecological Applications* 2: 203–217.

———. 1994. Biological inventory using target taxa: A case study of the butterflies of Madagascar. *Ecological Applications* 4: 407–422.

Kremen, C., R. K. Colwell, T. L. Erwin, D. D. Murphy, R. F. Noss, and M. A. Sanjayan. 1993. Terrestrial arthropod assemblages: Their use in conservation planning. *Conservation Biology* 8: 388–397.

Kremen, C., A. M. Merenlender, and D. D. Murphy. 1994. Ecological monitoring: A vital need for integrated conservation and development programs in the tropics. *Conservation Biology* 8: 388–397.

Lagos, N. A., and J. C. Castilla. 1997. Inferencia estadística bayesiana en ecología: Un ejemplo del uso en una reserva marina chilena. *Revista Chilena de Historia Natural* 70: 565–575.

Landres, P. B. 1983. Use of the guild concept in environmental impact assessment. *Environmental Management* 7: 393–398.

Landres, P. B., J. Verner, and J. W. Thomas. 1988. Ecological uses of vertebrate indicator species: A critique. *Conservation Biology* 2: 316–328.

Laurance, W. F. 1997. Effects of logging on wildlife in the tropics. *Conservation Biology* 11: 311–312.

Laurance, W. F., and R. O. Bierregaard Jr. 1997. *Tropical forest remnants: Ecology, management, and conservation of fragmented communities.* University of Chicago Press, Chicago, IL, USA.

Lawton, J. H., D. E. Bignell, B. Bolton, G. F. Bloemers, P. Eggleton, P. M. Hammond, M. Hodda, R. D. Holt, T. B. Larsen, N. A. Mawdsley, N. E. Stork, D. S. Srivastava, and A. D. Watt. 1998. Biodiversity inventories, indicator taxa and effects of habitat modification in tropical forest. *Nature* 391: 72–76.

Levey, D. J., and F. G. Stiles. 1992. Evolutionary precursors of long-distance migration: Resource availability and movement patterns in Neotropical landbirds. *American Naturalist* 140: 447–476.

Lewis, D. M. 1995. Importance of GIS to community-based management of wildlife: Lessons from Zambia. *Ecological Applications* 5: 861–871.

Likens, G. E., F. H. Bormann, N. M. Johnson, D. W. Fisher, and R. S. Pierce. 1977. *Biogeochemistry of a forested ecosystem*. Springer-Verlag, New York, NY, USA.

Lindenmayer, D. B., C. R. Margules, and D. B. Botkin. 2000. Indicators of biodiversity for ecologically sustainable forest management. *Conservation Biology* 14: 941–950.

Lodge, D. M. 1993. Biological invasions: Lessons for ecology. *Trends in Ecology and Evolution* 8: 133–137.

Ludwig, J. A., and J. F. Reynolds. 1988. *Statistical ecology*. John Wiley, New York, NY, USA.

Lugo, A. E. 1990. Removal of exotic organisms. *Conservation Biology* 4: 345.

Lyons, J., S. Navarro-Pérez, P. A. Cochran, E. Santana C., and M. Guzmán-Arroyo. 1995. Index of biotic integrity based on fish assemblages for the conservation of streams and rivers in west-central Mexico. *Conservation Biology* 9: 569–584.

Mabberley, D. J. 1988. The living past: Time state of the tropical rain forest. Pages 6–15 in E. R. C. Reynolds and F. B. Thompson, editors, *Forests, climate, and hydrology: Regional impacts*. United Nations University, Tokyo, Japan.

———. 1992. *Tropical rain forest ecology*, 2nd edition. Chapman and Hall, New York, NY, USA.

MacFarland, C. 1991. Goats on Alcedo Volcano in the Galápagos Islands: Help needed. *Conservation Biology* 5: 6–7.

Mack, R. N., D. Simberloff, W. M. Lonsdale, H. Evans, M. Clout, and F. A. Bazzaz. 2000. Biotic invasions: Causes, epidemiology, global consequences, and control. *Ecological Applications* 10: 689–710.

Magurran, A. E. 1988. *Ecological diversity and its measurement*. Princeton University Press, Princeton, NJ, USA.

Malcolm, J. R. 1994. Edge effects in central Amazonian forest fragments. *Ecology* 75: 2438–2445.

Mangel, M. et al. 1996. Principles for the conservation of wild living resources. *Ecological Applications* 6: 338–362.

Manly, B. F. J. 1986. *Multivariate statistical methods: A primer*. Chapman and Hall, London, UK.

———. 1991. *Randomization and Monte Carlo methods in biology*. Chapman and Hall, London, UK.

———. 1992. *The design and analysis of research studies*. Cambridge University Press, Cambridge, UK.

Mapstone, B. D. 1995. Scalable decision rules for environmental impact studies: Effect size, Type I, and Type II errors. *Ecological Applications* 5: 401–410.

Margoluis, R., and N. Salafsky. 1998. *Measures of success: Designing, managing, and monitoring conservation and development projects*. Island Press, Washington, DC, USA.

Martin, P. S., and C. R. Szuter. 1999. War zones and game sinks in Lewis and Clark's West. *Conservation Biology* 13: 36–45.

Mattoni, R., T. Longcore, and V. Novotny. 2000. Arthropod monitoring for fine-scale habitat analysis: A case study of the El Segundo sand dunes. *Environmental Management* 25: 445–452.

Mauchamp, A. 1997. Threats from alien plant species in the Galápagos Islands. *Conservation Biology* 11: 522–530.

McClanahan, T. R., and R. W. Wolfe. 1993. Accelerating forest succession in a fragmented landscape: The role of birds and perches. *Conservation Biology* 7: 279–288.

McCune, B. 1997. Influence of noisy environmental data on canonical correspondence analysis. *Ecology* 78: 2617–2623.

McGeoch, M. A. 1998. The selection, testing and application of terrestrial insects as bioindicators. *Biological Reviews* (Cambridge Philosophical Society) 73: 181–201.

McIntyre, S., and R. Hobbs. 1999. A framework for conceptualizing human effects on landscapes and its relevance to management and research models. *Conservation Biology* 13: 1282–1292.

McKenzie, D. H., D. E. Hyatt, and V. J. McDonald, editors. 1992. *Ecological indicators.* Elsevier Applied Science, London, UK.

McNeely, J. A. 1994. Lessons from the past: Forests and biodiversity. *Biodiversity and Conservation* 3: 3–20.

Mead, R. 1988. *The design of experiments: Statistical principles for practical application.* Cambridge University Press, Cambridge, UK.

Medellín, R. A., M. Equihua, and M. A. Amin. 2000. Bat diversity and abundance as indicators of disturbance in Neotropical rainforests. *Conservation Biology* 14: 1666–1675.

Meffe, G. K., and C. R. Carroll. 1997. *Principles of conservation biology,* 2nd edition. Sinauer Associates, Sunderland, MA, USA.

Merritt, R. W., and K. W. Cummins, editors. 1984. *An introduction to the aquatic insects of North America,* 2nd edition. Kendall/Hunt, Dubuque, IA, USA.

Mesquita, R. C. G., P. Delamônica, and W. F. Laurance. 1999. Effects of surrounding vegetation on edge-related tree mortality in Amazonian forest fragments. *Biological Conservation* 91: 129–134.

Miller, R. I., editor. 1994. *Mapping the diversity of nature.* Chapman and Hall, London, UK.

Miller, S. G., R. L. Knight, and C. K. Miller. 1998. Influence of recreational trails on breeding bird communities. *Ecological Applications* 8: 162–169.

Mills, L. S., M. E. Soulé, and D. Ï. Doak. 1993. The keystone-species concept in ecology and conservation. *BioScience* 43: 219–224.

Mittermaier, R. A., N. Myers, J. B. Thomsen, G. A. B. Da Fonseca, and S. Olivieri. 1998. Biodiversity hotspots and major tropical wilderness areas: Approaches to setting conservation priorities. *Conservation Biology* 12: 516–520.

Mitton, J. B., and M. C. Grant. 1996. Genetic variation and the natural history of quaking aspen. *BioScience* 46: 25–31.

Moguel, P., and V. M. Toledo. 1999. Biodiversity conservation in traditional coffee systems of Mexico. *Conservation Biology* 13: 11–21.

Murcia, C. A. 1995. Edge effects in fragmented forests: Implications for conservation. *Trends in Ecology and Evolution* 10: 58–62.

———. 1996. Forest fragmentation and the pollination of Neotropical plants. Pages 19–36 in J. Schelhas and R. Greenberg, editors, *Forest patches in tropical landscapes.* Island Press, Washington, DC, USA.

Murphy, D. D. 1990. Conservation biology and the scientific method. *Conservation Biology* 4: 203–204.

Murphy, K. R., and B. Myors. 1998. Statistical power analysis: A simple and general model for traditional and modern hypothesis tests. Lawrence Erlbaum Associates, Mahwah, NJ, USA.

The Nature Conservancy. 2000. *The five-S framework for site conservation: A practitioner's handbook for site conservation planning and measuring conservation success,* 2nd edition, Vols. I and II. The Nature Conservancy, Arlington, VA, USA.

Naughton-Treves, L. 1998. Predicting patterns of crop damage by wildlife around Kibale National Park, Uganda. *Conservation Biology* 12: 156–168.

Newman, J. A., J. Bergelson, and A. Grafen. 1997. Blocking factors and hypothesis tests in ecology: Is your statistics text wrong? *Ecology* 78: 1312–1320.

Nilsson, S. G., U. Arup, R. Baranowski, and S. Ekman. 1995. Tree-dependent lichens and beetles as indicators in conservation forests. *Conservation Biology* 9: 1208–1215.

Noss, R. F. 1990. Indicators for monitoring biodiversity: A hierarchical approach. *Conservation Biology* 4: 355–364.

———. 1991. *Biodiversity: Many scales and many concerns.* Proceedings of the Symposium on Biodiversity of Northwestern California, October 28–31, 1991, Santa Rosa, CA, USA, pages 17–22.

———. 1996. The naturalists are dying off. *Conservation Biology* 10: 1–3.

Novaro, A. J., K. H. Redford, and R. E. Bodmer. 2000. Effect of hunting in source-sink systems in the Neotropics. *Conservation Biology* 14: 713–721.

Osenberg, C. W., O. Sarnelle, S. D. Cooper, and R. D. Holt. 1999. Resolving ecological questions through meta-analysis: Goals, metrics, and models. *Ecology* 80: 1105–1117.

Osenberg, C. W., R. J. Schmitt, S. J. Holbrook, K. E. Abu-Saba, and A. R. Flegal. 1994. Detection of environmental impacts: Natural variability, effect size, and power analysis. *Ecological Applications* 4: 16–30.

Paoletti, M. G., editor. 1999. *Invertebrate biodiversity as bioindicators of sustainable landscapes*. Elsevier, New York, NY, USA.

Parsons, D. J., T. W. Swetnam, and N. L. Cristensen, editors. 1999. Uses and limitations of historical variability concepts in managing ecosystems. *Ecological Applications* 9: 1177–1277.

Pearson, D. L. 1994. Selecting indicator taxa for the quantitative assessment of biodiversity. *Philosophical Transactions of the Royal Society of London*, Series B 345: 75–79.

Pearson, D. L., and F. Cassola. 1992. World-wide species richness patterns of tiger beetles (Coleoptera: Cicindelidae): Indicator taxon for biodiversity and conservation studies. *Conservation Biology* 6: 373–391.

Perfecto, I., R. A. Rice, R. Greenberg, and M. E. Van der Voort. 1996. Shade coffee: A disappearing refuge for biodiversity. *BioScience* 46: 598–608.

Peterman, R. M. 1990. Statistical power analysis can improve fisheries research and management. *Canadian Journal of Fisheries and Aquatic Sciences* 47: 2–15.

Peters, R. H. 1991. *A critique for ecology*. Cambridge, University Press, Cambridge UK.

Peters, R. L., and T. Lovejoy, editors. 1992. *Global warming and biological diversity*. Yale University Press, New Haven, CT, USA.

Pickett, S. T. A., and P. S. White, editors. 1985. *The ecology of natural disturbance and patch dynamics*. Academic Press, New York, NY, USA.

Pielou, E. C. 1984. *The interpretation of ecological data*. John Wiley, New York, NY, USA.

Platt, W. J., and D. R. Strong Jr., editors. 1989. Gaps in forest ecology. *Ecology* 70: 535–576.

Poiani, K. A., B. D. Richter, M. G. Anderson, and H. G. Richter. 2000. Biodiversity conservation at multiple scales: Functional sites, landscapes, and networks. *BioScience* 50: 133–146.

Potvin, C., and D. A. Roff. 1993. Distribution-free and robust statistical methods: Viable alternatives to parametric statistics? *Ecology* 74: 1617–1628.

Powell, G. V. N., and R. Bjork. 1995. Implications of intratropical migration on reserve design: A case study using *Pharomachrus moccino*. *Conservation Biology* 9: 354–362.

Powell, G. V. N., J. Barborak, and M. Rodríguez S. 2000. Assessing representativeness of protected natural areas in Costa Rica for conserving biodiversity: A preliminary gap analysis. *Biological Conservation* 93: 35–41.

Powell, J. R., and J. P. Gibbs. 1995. A report from Galápagos. *Trends in Ecology and Evolution* 10: 260–263.

Power, A. G. 1996. Arthropod diversity in forest patches and agroecosystems of tropical landscapes. Pages 91–110 in J. Schelhas and R. Greenberg, editors, *Forest patches in tropical landscapes*. Island Press, Washington, DC, USA.

Power, M. E., D. Tilman, J. A. Estes, B. A. Menge, W. J. Bond, L. Í. Mills, G. Daily, J. C. Castilla, J. Lubchenco, and R. T. Paine. 1996. Challenges in the quest for keystones. *BioScience* 46: 609–620.

Primack, R., R. Rozzi, P. Feinsinger, R. Dirzo, and F. Massardo. 2001. *Elementos de conservación biológica: Perspectivas latinoamericanas*. Fondo de Cultura Económica, Mexico D.F., Mexico.

Putz, F. E. 1998. Halt the homogeocene. *Palmetto* 18: 7–9.

Putz, F. E., K. H. Redford, J. G. Robinson, R. Fimbel, and G. M. Blate. 2000. Biodiversity conservation in the context of tropical forest management. Environment Department Paper No. 75, Biodiversity Series, Impact Studies. World Bank, Washington, DC, USA.

Ralph, C. J., and J. M. Scott, editors. 1981. Estimating numbers of terrestrial birds. *Studies in Avian Biology* (Cooper Ornithological Society), no. 6.

Ralph, C. J., J. R. Sauer, and S. Droege, editors. 1995. *Monitoring bird populations by point counts*. General Technical Report PSW-GTR-149. Pacific Southwest Research Station, Forest Service, U.S. Department of Agriculture, Albany, CA, USA.

Reckhow, K. H. 1990. Bayesian inference in non-replicated ecological studies. *Ecology* 71: 2053–2059.

Redford, K. H. 1990. The ecologically noble savage. *Orion Nature Quarterly* 9(3): 24–29.

———. 1992. The empty forest. *BioScience* 42: 412–422.

Redford, K. H., and J. A. Mansour, editors. 1996. *Traditional peoples and biodiversity conservation in large tropical landscapes.* The Nature Conservancy, Arlington, VA, USA.

Reed, J. M., and A. R. Blaustein. 1995. Assessment of "nondeclining" amphibian populations using power analysis. *Conservation Biology* 9: 1299–1300.

Reid, R. S., and J. E. Ellis. 1995. Impacts of pastoralists on woodlands in South Turkana, Kenya: Livestock-mediated tree recruitment. *Ecological Applications* 5: 978–992.

Rice, R. E., R. E. Gullison, and J. W. Reid. 1997. Can sustainable management save tropical forests? *Scientific American* (April): 44–49.

Richardson, D. M. 1998. Forestry trees as invasive aliens. *Conservation Biology* 12: 18–26.

Ricketts, T. H., E. Dinerstein, D. M. Olson, and C. Loucks. 1999. Who's where in North America? *BioScience* 49: 369–381.

Riffell, S. K., K. J. Gutzwiller, and S. H. Anderson. 1996. Does repeated human intrusion cause cumulative declines in avian richness and abundance? *Ecological Applications* 6: 492–505.

Roberts, L. 1989. Disease and death in the New World. *Science* 246: 1245–1247.

Robinson, J. G., and E. L. Bennett. 1999. *Hunting for sustainability in tropical forests.* Columbia University Press, New York, NY, USA.

Rodríguez, J. P., D. L. Pearson, and R. Barrera R. 1998. A test for the adequacy of bioindicator taxa: Are tiger beetles (Coleoptera: Cicindelidae) appropriate indicators for monitoring the degradation of tropical forests in Venezuela? *Biological Conservation* 83: 69–76.

Rohlf, F. J., and R. R. Sokal. 1995. *Statistical tables,* 3rd edition. W. H. Freeman, San Francisco, CA, USA.

Roldán Pérez, G. 1992. *Fundamentos de limnología neotropical.* Editorial Universidad de Antioquia, Medellín, Colombia.

———. 1996. *Guía para el estudio de los macroinvertebrados acuáticos del Departamento de Antioquia.* Editorial Universidad de Antioquia, Medellín, Colombia.

Roosevelt, A. C. 1989. Resource management in Amazonia before the Conquest. *Economic Botany* 7: 30–62.

———. 1991. *Moundbuilders of the Amazon.* Academic Press, New York, NY, USA.

Rosenberg, D. M., and V. H. Resh, editors. 1993. *Freshwater biomonitoring and benthic macroinvertebrates.* Chapman and Hall, New York, NY, USA.

Rozzi, R., P. Feinsinger, and R. Riveros. 1997. *La enseñanza de la ecología en el entorno cotidiano.* Programa MECE Media, Ministerio de Educación, Santiago, Chile.

Rykken, J. J., D. E. Capen, and S. Mahabir. 1997. Ground beetles as indicators of land type diversity in the Green Mountains of Vermont. *Conservation Biology* 11: 69–78.

Saunders, D. A., R. J. Hobbs, and C. Margules. 1991. Biological consequences of ecosystem fragmentation: A review. *Conservation Biology* 3: 128–137.

Schelhas, J., and R. Greenberg. 1993. *Los fragmentos de bosques en el paisaje tropical y la conservación de las aves migratorias.* Migratory Bird Conservation Policy Paper no. 1. Smithsonian Migratory Bird Center, National Zoological Park, Washington, DC, USA.

———. editors. 1996. *Forest patches in tropical landscapes.* Island Press, Washington, DC, USA.

Schofield, E. K. 1989. Effects of introduced plants and animals on island vegetation: Examples from the Galápagos Archipelago. *Conservation Biology* 3: 227–239.

Schwartz, M. W. 1999. Choosing the appropriate scale of reserves for conservation. *Annual Review of Ecology and Systematics* 30: 83–108.

Severinghaus, W. D. 1981. Guild theory development as a mechanism for assessing environmental impact. *Environmental Management* 5: 187–190.

Shmida, A., and M. V. Wilson. 1985. Biological determinants of species diversity. *Journal of Biogeography* 12: 1–20.

Siegel, S., and N. J. Castellan. 1988. *Nonparametric statistics for the behavioral sciences,* 2nd edition. McGraw-Hill, New York, NY, USA.

———. 1995. *Estadística no paramétrica: Aplicada a las ciencias de la conducta.* Editorial Trillas, Mexico D.F., Mexico.

Simberloff, D. 1998. Flagships, umbrellas, and keystones: Is single-species management passé in the landscape era? *Biological Conservation* 83: 247–257.

Simberloff, D., D. C. Schmitz, and T. C. Brown. 1997. *Strangers in paradise: Impact and management of non-indigenous species in Florida.* Island Press, Washington, DC, USA.

Simberloff, D., J. A. Farr, J. Cox, and D. W. Mehlman. 1992. Movement corridors: Conservation bargains or poor investments? *Conservation Biology* 6: 493–504.

Simonetti, J. A. 1988. The carnivorous predatory guild of central Chile: A human-induced community trait? *Revista Chilena de Historia Natural* 61: 23–25.

Smith, S. M. 1995. Distribution-free and robust statistical methods: Viable alternatives to parametric statistics? *Ecology* 76: 1997–1998.

Sobel, D. 1995. Beyond ecophobia: Reclaiming the heart in nature education. *Orion* (autumn): 11–17.

Sokal, R. R., and C. A. Braumann. 1980. Significance tests for coefficients of variation and variability profiles. *Systematic Zoology* 29: 50–66.

Sokal, R. R., and F. J. Rohlf. 1995. *Biometry: The principles and practice of statistics in biological research,* 3rd edition. W. H. Freeman, San Francisco, CA, USA.

Soulé, M. 1990. The onslaught of alien species, and other challenges in the coming decades. *Conservation Biology* 4: 233–240.

Soulé, M. E., and G. Lease. 1995. *Reinventing nature?* Island Press, Washington, DC, USA.

Southgate, D., and H. L. Clark. 1993. Can conservation projects save biodiversity in South America? *Ambio* 22: 163–166.

Southwood, T. R. E. 1978. *Ecological methods,* 2nd edition. Chapman and Hall, London, UK.

Sparrow, H. R., T. D. Sisk, P. R. Ehrlich, and D. D. Murphy. 1994. Techniques and guidelines for monitoring Neotropical butterflies. *Conservation Biology* 8: 388–397.

Spellerberg, I. F. 1991. *Monitoring ecological change.* Cambridge University Press, Cambridge, UK.

Steel, R. G. D., and J. H. Torrie. 1960. *Principles and procedures of statistics, with special reference to the biological sciences,* 1st edition. McGraw-Hill, New York, NY, USA.

———. 1980. *Principles and procedures of statistics: A biometrical approach,* 2nd edition. McGraw-Hill, New York, NY, USA.

———. 1988. *Bioestadística: Principios y procedimientos,* 2nd edición. McGraw-Hill, Bogotá, Colombia.

Steidl, R. J., H. P. Hayes, and E. Schauber. 1997. Statistical power analysis in wildlife research. *Journal of Wildlife Management* 61: 270–279.

Steinberg, M. K. 1998. Neotropical kitchen gardens as a potential research landscape for conservation biologists. *Conservation Biology* 12: 1150–1152.

Stewart-Oaten, A. 1995. Rules and judgments in statistics: Three examples. *Ecology* 76: 2001–2009.

Stork, N. E., T. J. B. Boyle, V. Dale, H. Eeley, B. Finegan, M. Lawes, N. Manokaran, R. Prabhu, and J. Soberón. 1997. *Criteria and indicators for assessing the sustainability of forest management: Conservation of biodiversity.* Working Paper no. 17. Center for International Forestry Research (CIFOR), Bogor, Indonesia.

Strayer, D. L. 1999. Statistical power of presence-absence data to detect population declines. *Conservation Biology* 13: 1034–1038.

Struhsaker, T. T. 1997. *Ecology of an African rain forest: Logging in Kibale and the conflict between conservation and exploitation.* University Press of Florida, Gainesville, FL, USA.

Sutherland, W. J., editor. 1996. *Ecological census techniques.* Cambridge University Press, Cambridge, UK.

Taylor, B. L., and T. Gerrodette. 1993. The uses of statistical power in conservation biology: The vaquita and northern spotted owl. *Conservation Biology* 7: 489–500.

Temple, S. A. 1990. The nasty necessity: Eradicating exotics. *Conservation Biology* 4: 113–115.

Terborgh, J. 1999. *Requiem for nature*. Island Press, Washington, DC, USA.

Thiollay, J.-M. 1989. Area requirements for the conservation of rain forest raptors and game birds in French Guiana. *Conservation Biology* 3: 128–137.

———. 1992. Influence of selective logging on bird species diversity in a Guianan rain forest. *Conservation Biology* 6: 47–63.

———. 1995. The role of traditional agroforests in the conservation of rain forest bird diversity in Sumatra. *Conservation Biology* 9: 335–353.

Turner, I. M., and R. T. Corlett. 1996. The conservation value of small, isolated fragments of lowland tropical rain forest. *Trends in Ecology and Evolution* 11: 330–333.

Tyser, R. W., and C. A. Worley. 1992. Alien flora in grasslands adjacent to road and trail corridors in Glacier National Park, Montana (U.S.A.). *Conservation Biology* 6: 253–262.

Uhl, C. 1998. Perspectives on wildfire in the humid tropics. *Conservation Biology* 12: 942–943.

Uhl, C., P. Barreto, A. Veríssimo, E. Vidal, P. Amaral, A. C. Barros, C. Souza Jr., J. Johns, and J. Gerwing. 1997. Natural resource management in the Brazilian Amazon. *BioScience* 47: 160–168.

Umbanhower, C. E., Jr. 1994. Letter. *Bulletin of the Ecological Society of America* 74: 362–363.

Underwood, A. J. 1990. Experiments in ecology: Their logics, functions and interpretations. *Australian Journal of Ecology* 15: 365–389.

———. 1995. Ecological research and (research into) environmental management. *Ecological Applications* 5: 232–247.

———. 1997. *Experiments in ecology: Their logical design and interpretation using analysis of variance*. Cambridge University Press, Cambridge, UK.

van Shaik, C., J. W. Terborgh, and S. J. Wright. 1993. The phenology of tropical forests: Adaptive significance and consequences for primary consumers. *Annual Review of Ecology and Systematics* 24: 353–378.

Veblen, T. T., M. Mermoz, C. Martín, and T. Kitzberger. 1992. Ecological impacts of introduced animals in Nahuel Huapi National Park, Argentina. *Conservation Biology* 6: 71–83.

Verner, J. 1984. The guild concept applied to management of bird populations. *Environmental Management* 8: 1–14.

Vitousek, P. M. 1992. Global environmental change: An introduction. *Annual Review of Ecology and Systematics* 232: 1–14.

Vuilleumier, F., and M. Monasterio, editors. 1986. *High altitude tropical biogeography*. Oxford University Press, New York, NY, USA.

Wade, P. R. 2000. Bayesian methods in conservation biology. *Conservation Biology* 14: 1308–1316.

Walker, L. R., N. V. L. Brokaw, D. J. Lodge, and R. B. Waide, editors. 1991. Ecosystem, plant, and animal responses to hurricanes in the Caribbean. *Biotropica* 23 (special issue): 313–521.

Wallace, J. B., J. W. Grubaugh, and M. R. Whiles. 1996. Biotic indices and stream ecosystem processes: Results from an experimental study. *Ecological Applications* 6: 140–151.

Wallin, T. R., and C. P. Harden. 1996. Estimating trail-related soil erosion in the humid tropics: Jatun Sacha, Ecuador, and La Selva, Costa Rica. *Ambio* 25: 517–522.

Wells, M. P., and K. E. Brandon. 1993. The principles and practice of buffer zones and local participation in biodiversity conservation. *Ambio* 22: 157–162.

Wessman, C. A. 1992. Spatial scales and global change: Bridging the gap from plots to GCM grid cells. *Annual Review of Ecology and Systematics* 23: 175–200.

Western, D., and W. Henry, editors. 1994. Natural connections: Perspectives in community-based conservation. Oxford University Press, New York, NY, USA.

Whitmore, T. C., and G. T. Prance. 1987. *Biogeography and Quaternary history in tropical America*. Oxford University Press, New York, NY, USA.

Wiener, J. G., and M. H. Smith. 1972. Relative efficiencies of four small mammal traps. *Journal of Mammalogy* 53: 868–872.

Wiens, J. A. 1989. Spatial scaling in ecology. *Functional Ecology* 3: 385–397.

Wiens, J. A., and K. R. Parker. 1995. Analyzing the effects of accidental environmental impacts: Approaches and assumptions. *Ecological Applications* 5: 1069–1083.

Wilkinson, L., G. Blank, and C. Gruber. 1996. *Desktop data analysis with SYSTAT*. Prentice Hall, Upper Saddle River, NJ, USA.

Wilson, D. E., F. R. Cole, J. D. Nichols, R. Rudran, and M. S. Foster, editors. 1996. *Measuring and monitoring biological diversity: Standard methods for mammals*. Smithsonian Institution Press, Washington, DC USA.

Woodman, N., R. M. Timm, N. A. Slade, and T. J. Doonan. 1996. Comparison of traps and baits for censusing small mammals in Neotropical lowlands. *Journal of Mammalogy* 77: 274–281.

Wright, S. J., H. Zeballos, I. Domínguez, M. M. Gallardo, M. C. Moreno, and R. Ibánez. 2000. Poachers alter mammal abundance, seed dispersal, and seed predation in a Neotropical forest. *Conservation Biology* 14: 227–239.

Yoccoz, N. G. 1991. Use, overuse, and misuse of significance tests in evolutionary biology and ecology. *Bulletin of the Ecological Society of America* 72: 106–111.

Young, K. R. 1994. Roads and the environmental degradation of tropical montane forests. *Conservation Biology* 8: 972–976.

Zar, J. H. 1999. *Biostatistical analysis*, 4th edition. Prentice Hall, Upper Saddle River, NJ, USA.

Notes

Preface

1. The site conservation planning (SCP) framework, proposed by The Nature Conservancy, provides practical, step-by-step guidelines for developing effective conservation strategies at the landscape level and then evaluating the effect of those strategies. In addition to a comprehensive manual with a North American orientation (The Nature Conservancy 2000), The Nature Conservancy has produced other materials in English, Spanish, and Portuguese. The SCP framework and its objective guidelines can aid conservation efforts and decision making in any landscape—not just those landscapes where The Nature Conservancy and its partner organizations work. For more information, contact the International Site Conservation Program, Conservation Science Department, The Nature Conservancy, 4245 N. Fairfax Drive, Arlington, VA 22203, USA; or, by email, at: site_conservation@tnc.org.

Chapter 1

1. Similar definitions occur in Mangel (1996), Meffe and Carroll (1997), Schwartz (1999), Poiani et al. (2000), Primack et al. (2001), and the Site Conservation Planning manuals (The Nature Conservancy 2000).
2. For an excellent, far-reaching definition of management for conservation illustrated by case studies, see chapters 11–13 in Meffe and Carroll (1997), especially the section entitled "Four Basic Principles of Good Management," pages 352–359.
3. You'll find similar perspectives in, for example, Murphy (1990), Alpert (1995), Heinz (1995), Underwood (1995), and Castillo and Toledo (2000).

Chapter 2

1. The formal scientific method as presented in figure 2.1, clearly distinguishing scientific hypotheses from statistical hypotheses, resembles presentations by Fretwell (1972), Underwood (1990), and Peters (1991) but not that of Underwood (1997).

2. This real-life question was proposed during a workshop held in the Izozog region of the Gran Chaco of Bolivia, by parabiologists Alejandro Arambiza and Jorge Segundo, hunting monitor Odón Justiniano, and park guard Pedro Lizárraga.

3. This is the essence of "adaptive management" (Holling 1978) placed in the context of scientific inquiry. Such adaptive management, involving both professionals and community members, is the central theme of the key book by Margoluis and Salafsky (1998). Castillo and Toledo (2000) extend the approach to agroecology.

4. Kremen, Merenlender, and Murphy (1994) and Frumhoff (1996) provide excellent introductions to the philosophy of monitoring. Margoluis and Salafsky (1998) and the Site Conservation Planning manual (The Nature Conservancy 2000) provide social and regional contexts, respectively, for setting up monitoring projects. Note that monitoring in itself does not constitute a complete applied scientific inquiry, unless it meets certain criteria.

Chapter 3

1. In Latin America and elsewhere, the building of new roads may well be the greatest threat to the integrity of protected areas and even of the somewhat altered matrix outside. See Terborgh (1999) for an especially depressing picture. How many objective scientific inquiries, though, have actually dealt with the effects of roads or the consequences of alternative means of road construction and maintenance? Far more in the North Temperate Zone than in Latin America (although see Young 1994). See the review by Forman and Alexander (1998) as well as the "Special Section" in *Conservation Biology* 14(1) that begins with Hourdequin (2000).

Chapter 4

1. Numerous works deal with this and similar concerns about the effects of selective logging, and natural forest management overall, on conservation. Recent summaries or reviews include the following: Frumhoff (1995); Hartshorn (1995); Khan et al. (1995); Crome, Thomas, and Moore (1996); Laurance (1997); Rice, Gulleson, and Reid (1997); Uhl et al. (1997); Bawa and Seidler (1998); Chapman et al. (2000); Putz et al. (2000); and Fimbel, Grajal, and Robinson (2001). The subject is highly controversial, with proponents suggesting that natural forest management is the only viable mechanism to conserve forest outside of protected areas ("use it or lose it") and critics suggesting that the biota in the forests affected will only spiral downward ("use it *and* lose it"). The "use it or lose it" philosophy may be the most practical strategy to avoid complete, rapid forest degradation. On the other hand, recent books by two eminent field ecologists make a good if disheartening case for the "use it and lose it" scenario: Struhsaker (1997) and Terborgh (1999).

   At first glance, this question might not seem to comply with the criterion of "comparativeness." Implicit in the question, though, is the comparison between "no logging" and "logging." Indeed, any question that asks, "What is the effect of X?" automatically implies the phrase "as compared to the absence of X" and thus requires the inquiry's design to include a *control,* or situation where X does not occur. Examples include environmental impact studies and a vast number of questions important to conservation. While the detailed design of an environmental impact study is beyond the scope of this chapter, general considerations follow the scheme presented here. Additional perspectives on designing inquiries to assess environmental impacts include the following, among many other sources: Manly (1992), Osenberg et al. (1994), and Wiens and Parker (1995).

2. Instead of selective logging, the example in this chapter could have involved any one of many other themes critical to biodiversity conservation and protected area management in Latin America—or to basic field ecology. A few examples of such themes are: effects of fire and alternative strategies of fire management (e.g., Uhl 1998 and the set of papers that follows Uhl); effects of roads (see note 1, chap-

ter 3); effects of hunting on wildlife populations (e.g., Robinson and Bennett 1999); effects of visitor trails and how best to design them; or effects of the extraction of medicinal plants and other nontimber forest products. Other themes crop up in Meffe and Carroll (1997) and Primack et al. (2001).

3. Other books on study design, although they emphasize the nuances of statistical theory more than the realities of field research, may be of use to you. These include Mead (1988) and Manly (1992). Terminology and conceptual emphases in both books differ from those used here. Cook and Campbell (1979) employ yet another terminology—from the social sciences and the philosophy of science—but deal with many designs appropriate for our own field studies. I recommend Hurlbert (1984) as the best possible starting point if you wish to go further into design than just the scheme presented in this primer.

4. Again, different authors use different terms. I'll use experiment, experimental study, and manipulative study interchangeably, to describe situations where you (the investigator) have control over the events. In contrast, a nonexperimental, nonmanipulative, or observational study is one in which the events that the units of interest have experienced are out of your control or have occurred prior to your arrival on the scene. If you want to get technical, Eberhardt and Thomas (1991) present an exhaustive classification of studies and discuss the types of conclusions that can be drawn from each.

5. For example, with a design similar to 16 you could ask a complex question regarding differences between SL parcels and UL tracts, differences between seasons, and the possibility that the two contrasts aren't independent of one another—i.e., that the nature of differences between SL and UL sites changes with season (see also note 7 below). Here, "season" would be the second design factor, with two levels (wet season and dry season). The data analysis might be a bit tricky depending on the particulars of the design, and it's beyond the scope of this book.

6. Technically, a single statistical model, the General Linear Model, underlies the analysis of variance used for studies with categorical levels of the design factor, and the various regression techniques used to analyze studies with continuous levels. Of course, real statisticians have always known this, but in classic statistics texts written for basic or applied biologists, the two approaches have been presented separately. Only with the advent of powerful computers and versatile statistics programs has this relationship been hammered into us. Nevertheless, the practical use of the two approaches and the biological interpretation of results remain fundamentally distinct.

7. The texts listed in note 4 and the statistics texts listed in note 1 for chapter 5 provide full details on block designs. Dutilleul (1993) discusses the advantages that block designs have over randomized designs in field studies, while Newman, Bergelson, and Grafen (1997) present some crucial details of the statistical analysis—read both papers if you ever consider designing a study with blocks. Note that some, but not all, block designs provide you with three pieces of crucial information: (a) whether the design factor alone affects the values of the response variable, or the "treatment effect"; (b) whether the set of confounding factors that vary among blocks has directional effects on the values of the response variable, or the "block effect"; and (c) whether the relative effect of the design factor varies from block to block, or the "block X treatment interaction." For example, let's say that, unbeknownst to you, the cloud forest SL sites (figure 4.1) have fewer species of small mammals than nearby UL sites, whereas the opposite holds true for tropical dry forest. Such information might be crucial to the choice of management guidelines. In design 11 (figure 4.2e), though, that site-dependent switch in the difference between UL and SL sites would only contribute more variation to the data you collected from the mixed set of all replicate UL sites and the mixed set of all replicate SL sites, respectively. You wouldn't be able to discern any pattern through all that noise, and you probably couldn't conclude anything. A version of design 12 (figure 4.2f) with two evaluation units instead of one per response unit, however, would reveal the block X treatment interaction as well as the treatment effect alone and the block effect alone.

8. Excellent discussions on the consequences of selecting different quadrat sizes occur in Greig-Smith (1983), Ludwig and Reynolds (1988), and Krebs (1989). See also, though, the concise discussions in Sutherland (1996).

9. Some discussion of the choice of evaluation units for animals occurs in Southwood (1978), Ludwig and Reynolds (1988), Krebs (1989), and Sutherland (1996).

10. Wiener and Smith (1972), Krebs (1989), and Woodman et al. (1996) all discuss the fact that live traps for small mammals aren't always very good at sampling the target population.

11. These include in particular the texts listed in the three preceding notes. Numerous other books exist on sampling plants or particular animal groups, and many of them are excellent. Among recent entrants are three manuals produced by the Smithsonian Institution, Heyer et al. (1994) for amphibians, Wilson et al. (1996) for mammals, and Agosti et al. (2000) for ants.

12. Manly (1992) provides an in-depth discussion of optimal allocation of effort to replicate response units as opposed to multiple evaluation units (subsamples), as do the statistics texts cited in chapter 5, note 1.

13. See Hargrove and Pickering (1992). Some of the most important field studies in ecology, basic and applied, have involved only a single response unit per level of the design factor, but the interpretations are well justified nonetheless. For example, the most famous experiment ever performed on the effects of deforestation (Likens et al. 1977) in essence involved only one response unit of the control (a reasonably intact forest occupying one watershed) and one response unit of the experimental treatment (another watershed that was completely deforested). Nevertheless, little doubt exists that the deforestation itself was to blame for the tremendous contrasts in ecosystem function between the two sites that were then displayed.

14. Three key papers (Noss 1990; Kremen, Merenlender, and Murphy 1994; Frumhoff 1996) also bewail the lack of consistent definitions and rigorous design in monitoring for conservation purposes. Recent works that emphasize monitoring, such as Margoluis and Salafsky (1998), Poiani et al. (2000), and The Nature Conservancy (2000), present clear social and biological contexts for monitoring but do not evaluate the alternative definitions and designs of monitoring studies.

Chapter 5

1. Field biologists and conservation professionals tend to use three texts in particular: Steel and Torrie (1980, 1988), Sokal and Rohlf (1995), and Zar (1999). Each has its strengths, and we often find ourselves going back and forth among them. If you can obtain only one such text, get Zar. Note that the text by Sokal and Rohlf does not include statistical tables. You must obtain those separately (Rohlf and Sokal 1995). For a wonderful, irreverent introduction to classical statistics, read Gonick and Smith (1993). Once you think you know what you're doing but before you undertake your first statistical analysis, though, read the paper by Johnson (1999).

2. New statistical packages for desktop computers appear so frequently that I hesitate to list any here. Use what's available to you, what you like, or what a friend likes and can teach you to use. The section "Technological Tools" of the *Bulletin of the Ecological Society of America* is a great source for up-to-date reviews. As this chapter emphasizes, though, BEWARE! Statistical packages on computers are just sophisticated, sensitive, potentially dangerous tools that are far too easy to use, misuse, and abuse. If you don't know exactly what it is that the computer is doing to your data, if you don't know what the numbers generated actually mean, if you don't recognize the assumptions and biases of the particular statistical technique used or those of statistical inference in general, *don't even think about using a computer package.* If instead the calculations can be done with a simple hand calculator, do that and watch what happens at every step. If your analysis is far too complex for a hand calculator, it's always possible that your question and design are too complex as well.

3. *Parametric statistical inference* is technically the correct term, but most authors (including those of the leading texts) say *parametric statistics.*

4. In this imaginary example, the distribution of lengths of adult caimans doesn't deviate much from the theoretical "normal distribution" of values on which these procedures are based. Sizes of individuals in real-life animal populations, and many other response variables in field inquiries, often deviate much

more from the normal distribution. If each observation is transformed to its logarithm, though, the transformed (logarithmic) values often conform much more closely to the normal distribution than do the raw values. Therefore, Mead (1988) and some field biologists recommend that the raw data for any biological or ecological response variable, including but certainly not limited to size and weight of animals, be transformed to logarithms. Sample statistics calculated for transformed data, though, are difficult to interpret. Consequently, other biologists and statisticians urge caution, and so do I: *transform only if necessary, not automatically*. See Krebs (1989) for a clearly presented, practical discussion of the ways to transform field data and the need for doing that.

5. For these exercises, caimans were selected randomly from the list in table 5.1 by using the random numbers generator on my hand calculator. One rule of random sampling is that *every observation* (in this case, every caiman sampled and its length) *must be independent of preceding and succeeding observations*. Dra. Navideña (and we) must sample *with replacement*, that is, we must toss each caiman back into the lake after measuring it once, such that it has the same chance as any other caiman of being captured and measured again. If instead Dra. Navideña marks each caiman as she releases it and makes sure that she never measures it a second time, she's sampling *without replacement*. Where the statistical population of interest is not orders of magnitude larger than sample size n, as in this case, sampling without replacement will bias all the statistical procedures that follow unless you use special calculations developed for such cases. Where the statistical population of interest is considerably larger than sample size n, then it really doesn't matter whether you sample with or without replacement—but of course in this case there's very little chance of encountering the same individual twice, anyway! In short, if possible, *always sample with replacement*. If you can't do so, then view your statistical calculations with a bit of skepticism.

6. Technically, it's incorrect to express confidence limits by saying "the true value of the parameter we're estimating has an XX percent probability of lying between A and B." Such a straightforward definition belongs to a very different statistical approach, Bayesian statistics (discussed later in chapter 5). In the "classical" statistics that we're accustomed to using, to paraphrase Crome (1997) and others, the complete definition of confidence limits in this case would be as follows: If Dra. Navideña were to draw an infinite number of samples from the lake, each of n caimans, and calculate the 95 percent confidence interval each time, then 95 percent of those calculated intervals would include μ. We approximate this procedure by estimating the confidence interval based on a single sample, which gives us the interval from 2.0 m to 3.2 m inclusive. In reality, the true value of μ either does or doesn't lie within that interval; there's no "probability" involved. Nevertheless, because this technique usually gives the right answer (approximately 95 percent of the time in this case), Dra. Navideña chooses to believe that the true value falls between 2.0 m and 3.2 m. Phew! Now you realize why we need a simpler working definition.

7. See the texts mentioned in note 1 above, especially Sokal and Rohlf (1995). For the population CV, see Sokal and Braumann (1980). Calculations differ greatly for the different parameters. Be careful.

8. Important recent works that address this concern include the following: Noss (1991); Mapstone (1995); Crome (1997); and Steidl, Hayes, and Schauber (1997).

9. This relationship exemplifies *a priori* power analysis, the flip side of estimating the minimum n you need to achieve a given α and β with respect to the unknown δ. When doing an *a priori* power analysis, you ask, "What will be the β (or the power, $1 - \beta$) I'll achieve with a given sample size n, a given α_{rep}, and a given biologically significant difference δ?" It's very tempting to perform an *a posteriori* power analysis as well, on a set of data you'd already collected, asking, "What was my chance of having detected a given δ with these data?" or "What's the minimum δ I could have detected with these data?" For a number of esoteric reasons, an *a posteriori* power analysis is fraught with peril (Steidl, Nayes, and Schauber 1997; Gerard, Smith, and Weerakkody 1998), not to mention fraught with mind-boggling mathematics. Don't even think about it.

10. If you apply the method of appendix B to a set of preliminary data and find that it'll be impossible to achieve the desired power to evaluate your response variable and make reasonable decisions, don't despair (yet). Sometimes you can reconsider the way in which you were going to express the response

variable for the inquiry proper, and be able to achieve the desired power with reasonable sample sizes. For example, counts of earthworms per unit volume of soil (chapter 8) often vary wildly. Achieving reasonable statistical power to reveal biologically meaningful changes in earthworm abundance requires an astronomically large sample size and number of shovels. Biologically speaking, though, it makes sense to transform raw values for earthworm numbers to their logarithms and to be concerned about a δ between the logarithms of the counts at the different levels of the design factor instead of the δ between the raw counts themselves (also see note 4 above). Likewise, in a preliminary study the absolute count of seedlings of exotic plants might vary extremely widely from sample quadrat (the evaluation unit) to sample quadrat. You discover that it's impossible to achieve the desired power to evaluate the effect of adjacent land-use practices on the invasion of your reserve by these exotics, if this is to be your response variable. Try expressing the response variable not as the absolute counts but rather as the proportion of exotic seedlings among all seedlings (exotics + natives) together. Feed these values into the procedure in appendix B, and most likely you'll find that you can achieve adequate statistical power with a reasonable sample size. If so, use this as the response variable for the inquiry itself.

11. Besides the works listed in note 8 above and the technical references listed at the end of appendix B, see Gerrodette (1987), Peterman (1990), Taylor and Gerrodette (1993), Johnson (1999), and Strayer (1999). Also see the enlightening discussion that began with Reed and Blaustein (1995) and continued with a flurry of responses in *Conservation Biology* 11(1): 273–282 (1997).

12. The text on nonparametric statistics that is most widely used by field workers, especially in Latin America, is Siegel and Castellan (1988, 1995). Don't be put off by the title. Although all the examples involve the social sciences, a quick reading will assure you that they could as easily come from your own inquiries. There are frequent conceptual errors in the book. One example is the widespread promotion of one-tailed statistical tests. Use the tables with extreme caution. In general, though, the text is tremendously useful. Among other excellent texts on nonparametric statistics are Conover (1980) and Daniel (1990). Zar (1999) includes many of the most useful nonparametric tests alongside their parametric equivalents. Also see Potvin and Roff (1993) and some critical responses: Stewart-Oaten (1995), Smith (1995), and Johnson (1995).

13. Technically, the mathematics behind chi-square tests does involve parameters, but in the procedure itself you work directly with nominal data, and so chi-square tests are included with truly nonparametric tests in most textbooks.

14. Workers frequently apply either the Spearman rank correlation test or its better-known parametric analogue, the Pearson product-moment correlation, to a different class of designs that also have two observations per response unit. In studies with continuous levels of the design factor and one response variable, the investigator records the response unit's unique level of the design factor and the corresponding value of the response variable. *Don't ever apply a correlation test of any kind to such data.* (Having done so myself, I have the right to say this.) Correlation tests apply only to designs with two or more (in more complex cases) response variables, where the individual response units themselves constitute the design factor. In statistical terms, no obvious independent variable exists here, and there's no implication that changes in one measurement cause the changes in the other. For instance, a valid question might be, "Among the nations of the world, is there a relation between the frequency of earthquakes a nation experiences and the quantity of hot sauce consumed per year per inhabitant?" What's the design factor? Nations of the world. What are the two response variables? Frequency of earthquakes and per capita consumption of hot sauce. Would you say that one of those two is an independent variable that's likely to be the cause of changes in the other, dependent variable? *You* answer that one. What's the appropriate analysis? A correlation test, either Spearman or Pearson product-moment. In contrast, for the questions regarding the intensity of selective logging and the diversity and abundance of small mammals, forest birds, or forest frogs, intensity of selective logging is the design factor, the unique intensities experienced by different response units are the levels, and the measures of diversity and abundance are response variables. The appropriate analysis is regression, never correlation. The simplest parametric approach is *lin-*

ear regression. Unbeknownst to many investigators, there's also a nonparametric regression technique, which you'll find in the texts by Conover (1980) and Daniel (1990).

15. Log-linear analysis is discussed briefly by Sokal and Rohlf (1995). Agresti (1990) is a much more complete source but is quite overwhelming for nonstatisticians. The clearest and most practical presentation of log-linear analysis I've encountered is in Wilkinson, Blank, and Gruber (1996); despite the title, it does not require that you use the SYSTAT statistics package.

16. The many texts on multivariate statistics for conservation scientists and basic ecologists include Gauch (1982); Pielou (1984); Manly (1986); Digby and Kempton (1987); and Jongman, ter Braak, and van Tongeren (1995). The last may be the most useful to you. Don't forget about James and McCullough (1990), though. Those two authors developed some of the multivariate statistical techniques that are widely used today, and their cautionary notes (especially the last paragraph) should be heeded.

17. Sources include Manly (1991, 1992), Crowley (1992), and Gotelli and Graves (1996). Most of these deal exclusively with basic ecology and evolutionary biology, but the approaches can easily be applied to other fields.

18. See Fernández-Duque and Valeggia (1994), Arnqvist and Wooster (1995), Gurevitch and Hedges (1999), and Osenberg et al. (1999).

19. Wade (2000), who directly addresses the usefulness of Bayesian approaches to biological conservation, provides a good starting point. Follow this by reading the set of papers edited by Dixon and Ellison (1996). Several of these refer to comprehensive texts on the subject. The latter set of papers also includes at least one article that roundly criticizes the "rush to Bayesize" ecological and conservation-related inquiries and points out several advantages of classical over Bayesian statistics. Other key papers by Crome (1997) and Johnson (1999) briefly explain the approach behind Bayesian statistics. Also see Reckhow (1990) and one example of Bayesian inference regarding an applied question in a marine system (Lagos and Castilla 1997).

Chapter 6

1. Among numerous works on the independent migrations of plant and animal populations in response to past climate changes, the reshuffling of species assemblages time and time again, and implications to present-day and future conservation concerns are Graham (1986); Vuilleumier and Monasterio (1986); Whitmore and Prance (1987); Davis (1990); Foster, Schoonmaker, and Pickett (1990); and Meffe and Carroll (1997: 261–268).

2. The past decade has seen an explosion of papers, books, and research dollars devoted to the specter of climate change. Among the more straightforward sources, many of which address the problem of protected area management in the face of climate change, are Peters and Lovejoy (1992); Vitousek (1992); Kareiva, Kingsolver, and Huey (1993); Bawa and Markham (1995); Meffe and Carroll (1997: 153, 655–657); and Hughes (2000).

3. Many sources warn conservation professionals to take climate change into account when designing and managing protected areas. These include Graham (1988); Hunter, Jacobson, and Webb (1988); Halpin (1997); and Meffe and Carroll (1997: 333–335).

4. Despite abundant supporting evidence for this statement from throughout the Western Hemisphere (not to mention the Eastern Hemisphere, trodden upon by humans for much longer), some leading environmentalists choose to ignore it for fear that antienvironmentalists will use it to justify industrial-level exploitation of any and all landscapes. This fear is particularly prevalent in North America. See, for example, the debates in Soulé and Lease (1995). In my opinion, though, the profound influence of pre-Conquest peoples cannot logically be perverted into a justification for wholesale environmental destruction by today's "civilized" peoples. We must not blind ourselves to the influence of the past. For excellent discussions that emphasize that there's no such thing as a pristine landscape, see Hamburg and Sanford (1986), Mabberley (1988), Denevan (1992), Gómez-Pompa and Kaus (1992), and McNeely (1994).

5. The accumulating evidence for this proposal is discussed in Roberts (1989), Roosevelt (1989, 1991), and Gibbons (1990), among other sources.

6. See Mabberley (1988, 1992), and Gómez-Pompa and Kaus (1992).

7. Bucher (1987) and references therein address the case of the Chaco. For other dry tropical and subtropical habitats, see Janzen and Martin (1982), Janzen (1986a), and Simonetti (1988).

8. Graber (1995) presents an especially interesting perspective on this dilemma. I won't specifically address the issue of habitat restoration or "restoration ecology" in this book, because for most protected areas in Latin America and the Caribbean the cost of restoration would be prohibitive. If you're interested in the restoration philosophy, though, see Janzen (1995); Hunter (1996); Meffe and Carroll (1997: 479–511); the forum edited by Parsons, Swetnam and Cristensen (1999); and Primack et al. (2001: chapter 19).

9. If you're not already well aware of the fundamental role that small- and moderate-scale disturbance plays in tropical forests and other landscapes, see Pickett and White (1985), Denslow (1987), Platt and Strong (1989), Guariguata (1990), and Meffe and Carroll (1997: 243–245, 274–278, 309–310, and 316–320), among many other sources.

10. Because the "noble savage" myth profoundly influences management philosophy, policy, and politics from local to international scales (see the set of papers in *Conservation Biology* 14(5): 1351–1374 [2000]), I strongly suggest that you read Redford (1990). Also see Soulé and Lease (1995) and Redford and Mansour (1996).

11. For an excellent summary of the divergent opinions on exotic species management, see Meffe and Carroll (1997: 245–261). Also see Coblentz (1990), Lugo (1990), Soulé (1990), Temple (1990), Hobbs and Huenneke (1992), Lodge (1993), Cronk and Fuller (1995), Bright (1998), Devine (1998), and Mack et al. (2000).

12. See Schofield (1989), Powell and Gibbs (1995), and Mauchamp (1997).

13. Papers that discuss the spatial scales of different hummingbirds include Feinsinger (1976), Feinsinger and Colweil (1978), Feinsinger et al. (1988), and Levey and Stiles (1992). Few if any of your conservation concerns will involve hummingbirds, but they do provide great examples of diverse points of view on spatial scales.

14. See Shmida and Wilson (1985), Janzen (1986b), and Brown and Kodric-Brown (1977). Interestingly, these three different applications of "source-sink thinking" to biodiversity questions appeared before the explicit source-sink model itself.

15. Although their definition of sources and sinks differs somewhat from that employed here, Novaro, Redford, and Bodmer (2000) evaluate protected areas as "sources" for vertebrate species that are hunted outside the reserve, where the hunting pressure creates "sink" conditions.

Chapter 7

1. The field of landscape ecology has its roots in the semi-natural matrix (Brown, Curtin, and Braithwaite 2001) that covers much of Europe and the British Isles. There the highly visible mosaic of fields, pastures, woodlots, hedgerows, and other distinct habitat types long ago convinced many ecologists and land managers that no one habitat or "ecological community" should be considered independently of its neighbors. Forman and Godron (1986) authored a very readable text that emphasizes the European perspectives. Forman (1997), while still emphasizing the North Temperate Zone and Australia almost exclusively, gives an excellent, in-depth treatment. Many principles in both books apply to Latin American landscapes as well.

2. With apologies to Joni Mitchell, who wrote the song "Both Sides, Now," and to Judy Collins, who sang it more often.

3. See especially chapters 9 and 10 in Meffe and Carroll (1997). Two excellent, recent books explicitly

address habitat fragmentation in tropical landscapes: Schelhas and Greenberg (1996) and Laurance and Bierregaard (1997).

4. See, in addition to the sources cited in the preceding note, the following reviews: Saunders, Hobbs, and Margules (1991); Malcolm (1994); and Murcia (1995).

5. See Schelhas and Greenberg (1993, 1996) and Turner and Corlett (1996).

6. Key works that stress this point include Forman (1997); Gascon et al. (1999); McIntyre and Hobbs (1999); Dale and Haeuber (2000); Dale et al. (2000); Poiani et al. (2000); and Brown, Curtin, and Braithwaite (2001).

Chapter 8

1. I won't define *ecological integrity* for you. What do *you* think it means, with respect to your own landscape? Poiani et al. (2000) define it as "the ability to maintain component species and processes over long time frames." Karr (1996) defines virtually the same concept, *biological integrity,* in greater detail as "the capacity to support and maintain a balanced, integrated, adaptive biological system having the full range of elements (genes, species, assemblages) and processes (mutation, demography, biotic interactions, nutrient and energy processes, and metapopulation processes) expected in the natural habitat of a region." See also Karr's essay in Meffe and Carroll (1997: 483–485). Clearly, for a particular landscape, ecological integrity can be measured only in relative terms, against some baseline or control that represents 100 percent ecological integrity under that landscape's undisturbed conditions.

2. The term *ecological indicators* must be distinguished from another, similar-sounding term, *biodiversity indicators.* Biodiversity indicators are groups of animals (rarely plants) well known taxonomically, relatively easy to assess, and of broad geographic distribution. Their purpose is to help identify "biodiversity hot spots" for the biota in general, other concentrations of species diversity across many taxa, or other broad geographical patterns in species diversity. That is, ecologists and other conservation professionals expect that across the landscape, or across geographic regions, the number of species in the group selected as the biodiversity indicator will correlate with the number of species in many other taxa that are less easy to study (Pearson 1994; McGeoch 1998; Ricketts et al. 1999). Sometimes this may be wishful thinking (Lawton et al. 1998). In any event, note the difference in use between biodiversity indicators on the one hand, and ecological indicator species or species groups on the other hand. Even though the words *indicator* and *bioindicator* are often used without being further defined, it's imperative that you distinguish the two very different uses of the words and what they imply. Unfortunately, many authors and conservation professionals haven't done so, and this has led to confusion.

3. Key resources that discuss ecological indicators in general include Noss (1990), an essay by Noss in Meffe and Carroll (1997: 88–89), and Kremen et al. (1993). You'll find exhaustive lists of indicator, and target, variables of all kinds in Spellerberg (1991) and McKenzie, Hyatt, and McDonald (1992). To monitor many of the variables that the last two books present, you'll need advanced technology, specialized technicians, and lots of money. I suggest you turn to Stork et al. (1997) and Lindenmayer, Margules, and Botkin (2000), although even in these thoughtful works several of the ecological indicators suggested may be beyond your reach for financial or practical reasons.

4. This scheme generally follows that given in Meffe and Carroll (1997: 69–70) as well as schemes, and doubts, presented by Simberloff (1998), Caro and O'Doherty (1999), Schwartz (1999), and Poiani et al. (2000).

5. Power et al. point out that the keystone species concept has great value in basic ecology. Indiscriminate use of the label "keystone" in conservation and management, though, poses risks (Mills, Soulé, and Doak 1993). In contrast to a recent paper (Simberloff 1998) that's highly influential in management circles, I'll propose that the overuse of the label "keystone," and the unanticipated risks it incurs, aren't adequately recognized. See Kotliar (2000) and also Poiani et al. (2000).

6. At least four key references present lists of criteria for indicator groups or species. Favila and Halffter (1997), Hilty and Merenlender (2000), and Caro and O'Doherty (1999) present excellent general schemes. McGeoch (1998) offers an especially complete, rigorous, and objective means for evaluating potential ecological indicators.

7. Some time ago Severinghaus (1981), Landres (1983), and Verner (1984) proposed that the guild profile of the avifauna might serve as an ecological indicator. Recently some doubts have arisen (e.g., Poiani et al. 2000), but the use of ecological indicators based on guilds—and not only guilds of birds—still holds great promise in tropical and subtropical regions (Stork et al. 1997).

8. See Kremen (1992, 1994), Sparrow et al. (1994), and Stork et al. (1997).

9. Among many works on dung beetles as a target or indicator group, see especially Klein (1989); Hanski and Cambefort (1991); Halffter, Favila, and Halffter (1992); Halffter and Favila (1993); and Favila and Halffter (1997).

10. See Kremen et al. (1993), Didham et al. (1996), and Hilty and Merenlender (2000). Mattoni, Longcore, and Novotny (2000) used a great variety of insects as ecological indicators of the success of efforts at restoration. Chapters in Paoletti, ed. (1999), address insects as indicators of the "ecological sustainability" of different farming practices.

11. If this paragraph doesn't convince you that worms might make an excellent choice as an indicator group for the ecological integrity of soils, an important target group, and a group appreciated by local people all at the same time, read Darwin (1881) and then Fragoso (1989, 1992); Fragoso et al. (1993); Hendrix (1995); or González, Zou, and Borges (1996).

12. The voluminous literature on aquatic invertebrates, many sources also providing data on sensitivity to pollution, includes numerous books of fundamental importance. Among these are Hurlbert (1981), Merritt and Cummins (1984), and Roldán Pérez (1992, 1996). Merritt and Cummins focus on the United States and Canada. Much of their information is also valid for Latin America, where every group of conservation professionals should also have a copy of Roldán Pérez (1992) and refer to it often. The reviews in Rosenberg and Resh (1993) will convince you that benthic insects provide an extraordinarily powerful tool for monitoring or investigating the ecological integrity of the aquatic environment. Finally, general books on ecological techniques often include sampling methodologies for benthic insects. See, for example, Southwood (1978) and Sutherland (1996).

13. Any one of the following four sources presents the philosophy and practical development of the Benthic Index of Biological Integrity: Kerans and Karr (1994); Fore, Karr, and Wisseman (1996); or Karr and Chu (1997, 1998). Also see appendix C for a more direct means of learning about this methodology.

14. Trout from the North Temperate zone were introduced into cool-water South and Central American streams and lakes long before anyone worried about effects on the native fauna. In my opinion these potentially serious effects should generate many conservation concerns and subsequent inquiries. The few related studies performed to date—for example, about the effects of introduced salmon on aquatic environments in southern Chile—are cause for nightmares, as is one study from New Zealand (Flecker and Townsend 1994).

15. Buchmann and Nabhan (1996) and Kearns, Inouye, and Waser (1998) point out that the plant-pollinator interaction is critically important both as a target and as an ecological indicator. Several studies (e.g., Aizen and Feinsinger 1994; Murcia 1996) have focused on the interaction in its own right, examining rates of pollination and fruit production with respect to habitat fragmentation or distance from the forest-agriculture boundary. Nevertheless, you could easily reverse the reasoning and exploit pollination levels or frequencies of fruits produced as indicators of ecological integrity overall. Dafni (1992) and Kearns and Inouye (1993) describe techniques for investigating just about any of the numerous possibilities for response variables.

Chapter 9

1. Hundreds of texts and journal articles present techniques for quantifying species diversity. Although the philosophy and recommendations differ somewhat from those presented here, Krebs (1989) presents an exceptionally clear and well-presented survey. Magurran (1988) provides an even more complete compendium of alternatives. The second book's philosophy differs quite a bit from that presented here, and it's probably most useful for questions in basic and theoretical ecology.

2. If you need to compare the S of samples that differ greatly in the sampling effort employed or the number of individuals encountered, you might wish to use the estimates produced by any one of various "rarefaction" techniques. These mathematical techniques provide an estimate of what S would be if all samples were the same size and/or represented the same effort. One technique presented by Hurlbert (1971) employs a rarefaction index. Krebs (1989) and Colwell and Coddington (1994) further discuss rarefaction approaches.

3. Originally, ecologists used rank-abundance graphs to compare observed patterns of species abundance distributions with those predicted by mathematical models, such as the log-normal distribution, the geometric series or log series distributions, and the so-called broken-stick model. Fortunately, most of those models—except for the log-normal distribution, which is always valid—and their labored ecological explanations have been discarded. Many field ecologists and conservation professionals are beginning to recognize just how valuable these empirical graphs are in themselves.

Chapter 10

1. In line with the orientation of the rest of this book, chapter 10 intentionally avoids addressing the exceedingly important social, political, and economic dimensions of "participatory management" and "community-based conservation." For full discussions of these and related themes, see for example Southgate and Clark (1993), Wells and Brandon (1993), Western and Henry (1994), Margoluis and Salafsky (1998), and Hackel (1999).

Index

About the Author

Many years ago Peter Feinsinger confused "Neotropical" with "Nearctic" while applying to the doctoral program at Cornell University. Being too embarrassed to admit to his mistake, he abandoned the idea of research in the North Temperate zone and became a tropical ecologist (later adding the southern subtropical and temperate regions as well). Since 1971 he has worked in many landscapes in Central and South America and the Caribbean, first performing research on themes in basic ecology and more recently participating in projects in conservation ecology, training, and public education. In 1992 he retired as professor of zoology and courtesy professor of Latin American studies at the University of Florida. Currently, he resides in Flagstaff, Arizona, where he's adjunct professor of biology at Northern Arizona University and long-distance conservation fellow with the Wildlife Conservation Society. Often, though, he's working with any one of a number of local, national, or international institutions in South America.